Designer Drugs Directory

Designer Drugs Directory

Authors:

Karel Valter and Philippe Arrizabalaga

Service de l'Ecotoxilogie Cantonal, Genève

Directeur: Dr. Jean Claude Landry

ELSEVIER

AMSTERDAM – LAUSANNE – NEW YORK – OXFORD – SHANNON – TOKYO 1998

ELSEVIER SCIENCE S.A.
Avenue de la Gare 50
1001 Lausanne 1, Switzerland

Library of Congress Cataloging in Publication Data
A catalog record from the Library of Congress has been applied for.

ISBN: 0-444-20525-X

Transferred to digital printing 2005.

Contents

Preface

This book is intended to serve as a quick reference handbook on so called designer drugs. These new, mainly synthetic compounds are also often referred to as analogues of controlled substances. These powerful drugs represent a considerable danger to public health because they are usually easily prepared from readily available, inexpensive chemicals in clandestine laboratories and marketed by unscrupulous dealers.

This class of drugs of abuse is rapidly growing in variety and number of the compounds. Careful analysis shows that about 1000 potent substances can be reasonably prepared in an average clandestine laboratory and this number rises to a potential of several thousands of different drugs if a use of more sophisticated chemical procedures is taken in account.

Fortunately, only a fraction of this enormous potential has appeared in the illicit traffic. Identification, toxicology and other properties have been thoroughly investigated for several compounds, known for their widespread abuse but their analogues and derivatives remain poorly documented. Moreover, the data on these compounds are scattered in the literature so that they are not rapidly available. A huge amount of data on designer drugs, becoming often from clandestine sources can also be found on the Internet. A critical overview of all these facts has become urgent.

Consequently, the book provides the essential information on 107 designer drugs which is separated in two main sections.

The first part describes briefly some aspects of designer drugs manufacture (its advantages, and its geographic distribution), new trends of their abuse, sources of information on these substances and the employed terminology in this book.

In the second, descriptive section, these drugs are classified into ten main categories according to their chemical structures and their prevalent pharmacological action. Each category of the designer drugs is described in a separate chapter. Each chapter is then followed by a set of corresponding data sheets providing basic data on each particular drug (its computer generated IUPAC chemical name, using Autonom 1.1 software, chemical structure, Chemical Abstracts registry number, Chemical Abstracts chemical name) together with its street names and eventual synonyms. Basic toxicological data (human active dose, duration and type of action, toxic effects and toxicity), short notes on history of a particular drug as well as the most pertinent bibliographic references are also included, if available.

The book is thoroughly indexed. Along with the usual Subject Index it contains the Street Names Index, listing more than 230 street names of the described designer drugs. Of course, a list of employed abbreviations and a short glossary are also provided.

Due to the rapid evolution in this field, this directory can hardly be exhaustive, but a considerable effort has been made to make it as complete as possible.

I hope that the reader will find this book helpful and easy to use.

Jean-Claude Landry
Director of the Institute of Ecotoxicology (ECOTOX),
Geneva, Switzerland

1 Introduction

1.1 Definition of Designer Drugs

A rapidly growing number of synthetic drugs have appeared on the clandestine drug scene in the last decade. The term « designer drug » was coined for this class of drugs of abuse in 1985 [1]. According to Henderson [2], this term should be reserved only for those drugs which are synthesised from readily available precursors, marketed under attractive « trade-marks » and not subjected to legal control as substances of abuse.

This recommendation has not been followed and the term is actually applied to practically all synthetic drugs of clandestine origin.

These new drugs are also referred to as controlled substances analogues. The latter designation is of slightly narrower scope because there are several designer drugs which cannot be reasonably considered as analogues of any controlled substance.

1.2 Terminology

Designer drugs can be classified into several categories according to their chemical structure and their prevalent pharmacological action.

While their chemical classification is rather straightforward, the pharmacological classification is more complicated due to the multiplicity of pharmacological categories. Thus, there are about 10 different terms only for drugs commonly known as hallucinogens.

For this reason, the names of the pharmacological classes used in this publication (in bold) are listed here along with their sometimes very approximative synonyms.

Psychotomimetics (psychedelics, hallucinogens, psychodysleptics, eidetics, phantastica, psychotogens, entheogens). The scope of this term (meaning « psychosis mimicking ») is larger than that of hallucinogens, because many substances in this class do not produce hallucinations in man.

Consequently, if a given psychotomimetic also shows pronounced hallucinogenic action in man, this property will be further defined as follows: Psychotomimetic, hallucinogen.

Deliriants (centrally active anticholinergics, hallucinogenic glycolates, atropine-like hallucinogens)

CNS Stimulants (psychoanaleptics, psychostimulants)

1.3 New Trends of Drug Abuse

Rapid evolution in the field of designer drugs has brought about new behaviour patterns in drug abuse touching many levels of society.

Thus, a particular drug is absorbed as an « aesthetic enhancer » to improve visual and auditive perception before going to a concert, a theatre or an exhibition of paintings. In this case the most frequently abused drugs are psychotomimetic phenethylamines such as 2,5-dimethoxy-4-bromophenethylamine (2CB) or 2,5-dimethoxy-4-methylamphetamine (DOM) in small doses. Another type of drug is taken during « Rave parties » to produce euphoria, enhance empathy and communication among participants, particularly 3,4-methylenedioxymethamphetamine (MDMA), its derivatives and etryptamine. There are also several drugs which are absorbed before making love because of stimulation of sexual pleasure and desire (2CB, 2CI) or removing inhibitions, such as metaqualone, its analogues and sodium oxybate (GHB).

Use of several designer drugs (MDA, MDMA in small doses) to promote meditation, particularly in Zen Buddhism, has also been reported [3]. Some of these drugs even initiated the creation of various philosophic trends (e.g. Ketamine).

Designer drugs have also been misused in a very irresponsible way to intoxicate innocent participants of a party or a concert. In these cases, the drugs (psychotomimetics, phencyclidines or deliriants) are usually added to various soft drinks [4].

The mode of administration of these drugs has also considerably evolved. Although the drug is still mostly injected, swallowed or drunk in a solution, intranasal, sublingual and transdermal (particularly for opiates) administrations are also current.

Absorption of a mixture of two or even more drugs, known for their synergism, seems more and more popular. These synergetic substances may also be taken in a precise time interval to enhance or modify the action of these drugs (« priming », see Glossary).

For instance, a dose of MDMA as the « primer » is followed by a dose of 2CB to obtain a new and unpredictable effect.

Similarly, a small dose of a drug may serve as a « booster », if it is absorbed shortly after the absorption of the principal portion of the same drug. This is a frequent way to prolong the effect of a relatively short acting drug (e.g. MDMA).

Pre-treatment by inhibitors of monoamine oxidase, usually for several days before a drug intake to amplify the drug's action, is also a quite frequent, very hazardous and potentially lethal practice.

Often, a true cocktail of drugs is absorbed, particularly due to the presence of drug mixtures in the clandestine market. These mixtures are produced to replace a drug which is momentarily out of stock.
When in turn a consumer takes such a mixture with another drug, serious and complicated intoxication may occur.

1.4 Geographic Distribution of Abuse and Manufacture of the Designer Drugs

About twelve years ago the problem of designer drugs was practically limited to the US territory.

Today, the situation is very different, particularly due to political changes in the world in the late 1980's. Practically all countries of Europe, Australia and Canada are confronted with the problem of abuse and manufacture of designer drugs [5].

New centres of clandestine production of these drugs have appeared in the countries of Eastern Europe, mainly due to the absence of control of various precursors, insufficient legislature, chaotic industry and unemployment of qualified chemists.

Interestingly, clandestine laboratories in Hungary, Czech Republic, Lithuania and Estonia seem to specialise in the production of amphetamines while those in Russia produce synthetic opiates such as 3-methylfentanyl [5].

Thus, 70 amphetamine producing clandestine laboratories were discovered in the Czech Republic in 1994. Some of them were under Dutch financial control and produced MDMA for the Dutch and German markets. A similar case, involving 50 kg of MDEA, was recently reported in Hungary [6].

1.5 Estimation of the Potential Number of Designer Drugs

This class of drugs of abuse is rapidly growing in the variety and number of compounds. Careful analysis, based on the structure-activity relationships and the technical feasibility shows that at least 1000 potent substances can be reasonably easily prepared in an average clandestine laboratory.

This number rises to a potential of several thousands of different drugs if the use of more sophisticated chemical procedures is taken into account.

Indeed, recent research in the USA has shown that these techniques are no longer out of reach of a clandestine manufacturer. The estimated potential number of powerful designer drugs in each category is given below.

Psychotomimetic phenethylamines 250

Psychotomimetic indolealkylamines 250-300

LSD analogues 10

Synthetic canabinoids 10

Phencyclidines 50

Deliriants 50

CNS stimulants 100

Opiates 500-4000 (1400 fentanyls)

1.6 Sources of Information about Designer Drugs

The open literature represents only a part of the information on these substances. The rapid evolution in the field of clandestine drug abuse and manufacture can be also observed on the Internet where a great number of sites contains pages on various designer drugs.

An intense exchange of information concerning these drugs in several specialised Newsgroups [7] is also very interesting.

It is almost impossible to list all the addresses of these sites, particularly because of their frequent change. Fortunately, today they can be found and updated easily using the extremely powerful searching computer at Alta Vista of Digital (http://www.altavista.digital.com)

Various publications about clandestine manufacture of different drugs of abuse have also been consulted.

Of course, these data, coming mainly from clandestine sources, have been treated with due precaution.

In this case, a particular designer drug has been taken into consideration only if its behavioural action in man, showing a high abuse potential, and its easy synthesis were clearly described. In addition, the description has been compared to the data obtained from other sources. Several such substances have been considered as « emerging » designer drugs and included in the directory. Also, in spite of only limited evidence of their abuse, a couple of compounds found in the open literature and showing a particularly high abuse potential have also been included in the text.

1.7 General References

[1] Baum, R.: Chemical and Engineering News, **9**, pp.7-16, (1985)
[2] Henderson, G.L.: J. Forensic Sci., **33**, p.569, (1988)
[3] Watson, L.; Beck, J.: J. of Psychoactive Drugs, **23**(3), p.261, (1991)
[4] Ragan, F.A.; Hite, S.A.; Samuels, M.S.; Garey, R.E.: J. Analyt. Toxicol., **9**, p.91, (1985)
[5] Report of the International Narcotic Control Board, (1995)
[6] Korosi, A.; Nagy, J; Nagy, G.; Gal, T.; Veress, T. Microgram, XXVII, (2), p.21, (1994)
[7] Newsgroups: alt.consciousness; alt.consciousness.mysticism; alt.consciousness.near-death-exp; alt.culture.zippies; alt.drugs; alt.drugs.chemistry; alt.drugs.culture; alt.drugs.hard; alt.drugs.pot; alt.drugs.pot.cultivation alt.drugs.psychedelics; alt.hemp; alt.hemp.politics; alt.hemp.recreational; alt.law-enforcement; alt.personals.psychedelic; alt.psychoactives; alt.religion.shamanism; bionet.neuroscience; clari.news.drugs; fido7.drugs; rec.drugs.cannabis; rec.drugs.chemistry; rec.drugs.misc; rec.drugs.psychedelics; sdnet.hemp

2 Description of Designer Drugs

The designer drugs can be classified, according to their chemical structures and pharmacological profiles, into the following categories:

2.1 Psychotomimetic phenethylamines
2.2 LSD analogues
2.3 Psychotomimetic indolealkylamines
2.4 Synthetic cannabinoids
2.5 Phencyclidine and its congeners
2.6 Deliriants
2.7 CNS Stimulants
2.8 Synthetic opiates
2.9 Metaqualone and its analogues
2.10 GHB

In the following chapters the most important designer drugs are described and their active doses (as hydrochloride salts, if not specified otherwise) in non-tolerant man are given, if available. These data have been obtained from all accessible literature and they are presented here only for information.

More detailed facts, concerning each particular drug, can be found in the data sheets attached to each chapter. The toxic manifestations listed in the data sheets are those which are likely to be observed in an overdose by a particular drug.

2.1 Psychotomimetic Phenethylamines

The majority of the psychotomimetic phenethylamines has been synthesised since 1960. More than a hundred of these substances have been tested and found active in humans [1,2]. Most of them may be prepared relatively easily in an average chemical laboratory. Hence, they represent synthetic drugs of choice to a clandestine manufacturer.
This class of drugs may be further divided into two subgroups.

The first subgroup contains phenethylamines having, in man, communication enhancing and euphoriant properties. Their effects are practically free of visual distortions and hallucinations. This subgroup has received particular attention here because it includes 3,4-methylenedioxymetamphetamine (MDMA), its derivatives and their substitutes which are the most important drugs of abuse among psychotomimetic phenethylamines.

The second subgroup includes phenethylamines showing a human pharmacological profile similar to mescaline (but up to 400 times more potent) with prominent visual illusions, hallucinations, altered perceptions of body image, colours, sounds, space and time. Further psychic effects may include euphoria, depersonalisation, emotional lability, anxiety, fear, hostility and impaired judgement. These substances are usually referred to as classical hallucinogens.

This pharmacological class is still evolving and new psychotomimetic phenethylamines appear in the literature. Several excellent reviews on their chemistry and pharmacology have been published [1,3,4,5]. Anyone interested in phenethylamines must read a remarkable book [2] on these compounds since one of its authors is one of the world's authorities in psychotomimetics.

2.1.0 General References:

[1] Glennon, R.A.: Classical Hallucinogens. In: Pharmacological Aspects of Drug Dependence, Handb. Exp. Pharm., **118**; Eds.: Schuster, C.R.; Kubar, M.J.; Springer Verlag, pp.343-371, (1995)
[2] Shulgin, A.T. and Shulgin, A.: PIHKAL, Transform Press, Inc., P.O.Box 13675, Berkeley, CA (1991)
[3] Shulgin, A.T.: Psychotomimetic drugs: Structure-Activity Relationships. In: Handbook of Psychopharmacology, Eds. Iversen, L.L.; Iversen, S.D.; Snyder, S.H., Plenum Press, New York, pp.243-333, (1978)
[4] Clare, B.W.: J. Med. Chem., **33**, p.687, (1990)
[5] Nichols, D.E.: Medicinal Chemistry and Structure-Activity Relationships of Amphetamines. In: Amphetamine and its analogs; Cho, A.K. and Segal, D.S. (eds.); Academic Press, pp.3-41, (Chapter 1), (1994)

2.1.1 MDMA and its Derivatives

This chapter describes MDMA and those closely related MDMA analogues possessing nearly identical psychotomimetic properties in man.

The world-wide and still increasing abuse of MDMA is due to its euphoriant properties and capacity to enhance communication and contact with other people. This last property is particular to a small class of compounds called « entactogens » or « empathogens » (meaning « creating contact or empathy ») [1].

In man, these compounds also frequently produce physiological effects such as cardiovascular disturbances (tachycardia, hypertension, arrhythmia), hyperthermia and dehydration which may sometimes be fatal. In animal studies, after a repeated administration, some of these substances have been demonstrated to be neurotoxic, causing practically permanent lesions of serotoninergic neurons in the brain of treated animals [2].

Use of these substances by young people has given rise to a new sub-culture « Rave » and to a particular way of spiritual development (« Lion Path ») [9].

A huge amount of information on these phenethylamines is found in the scientific literature, on the Internet and also in various publications of mostly clandestine origin.

Of the number of scientific publications on MDMA and its derivatives, there are excellent reviews on their neurotoxicity [2], abuse [3] and clandestine manufacture [4]. Many pages on MDMA [5] and its clandestine synthesis [6,7] can be found on the Internet. Also, there are four comprehensive manuals on the utilisation of these phenethylamines including chapters on safe dancing under the influence of these drugs [8]. The complete text of one manual can be downloaded [9].

3,4-Methylenedioxyamphetamine (**1**, MDA, « Love drug », « Zen ») is the first member of this class. It was a very popular drug of abuse in the late 1960s.

Modification of its chemical structure has given MDMA (« ADAM », « Ecstasy », « XTC ») (**2**, 3,4-methylenedioxymethamphetamine), MDEA

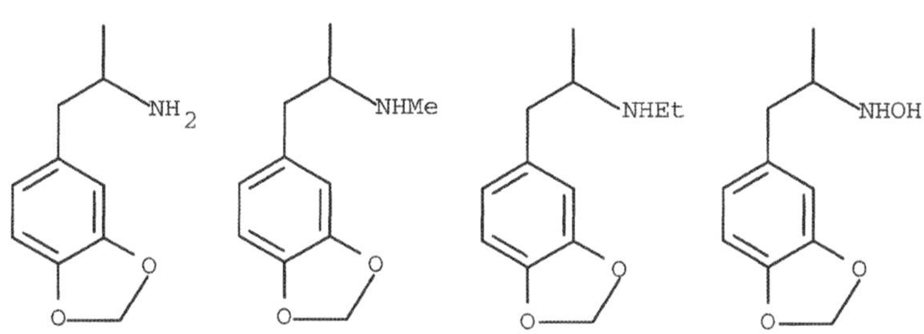

1	**2**	**3**	**4**
80-150 mg	80-200 mg	100-200 mg	80-200 mg

(« EVE », **3**, N-ethyl-3,4-methylenedioxyamphetamine) and MDOH (**4**, N-hydroxy-3,4-methylenedioxyamphetamine). All these compounds possess entactogenic properties [3] but those of MDMA are the most pronounced.

The N-propyl derivative of MDA, MDPR (**5**, 3,4-methylenedioxy-N-propylamphetamine), does not exhibit any properties similar to MDMA but there is an evidence of its clandestine manufacture.

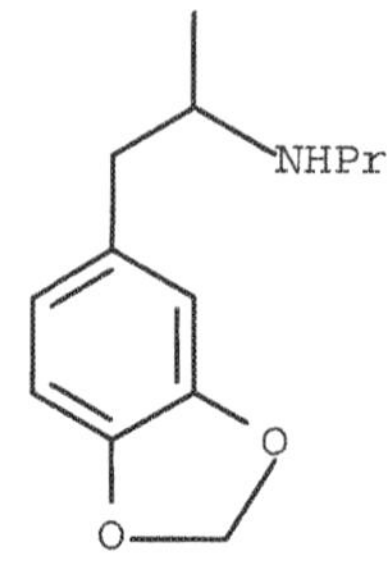

5

150-200 mg

2.1.1.1 General References

[1] Nichols, D.E.; Hoffman, A.J.; Oberlender, R.A.; Jacob, Peyton III; Shulgin, A.T.: J. Med. Chem., **29**, p.2009, (1986)
[2] Gibb, J.W.; Hanson, G.R.; Johnson, M.: Neurochemical Mechanisms of Toxicity. In: Amphetamine and its analogs; Cho, A.K. and Segal, D.S. (editors); Academic Press, pp.269-282, (Chapter 9), (1994)
[3] McCann, Una; Ricaurte, G.A.: Use and Abuse of Ring Substituted Amphetamines. In Amphetamine and its analogs; Cho, A.K. and Segal, D.S. (editors); Academic Press, pp.371-381, (Chapter 12), (1994)
[4] Del Cason, T.A.: J. Forensic Sci., **35**, p.675, (1990)
[5] http://www.ecstasy.org and http:// hyperreal.com/drugs/mdma
[6] http://www.links.net/drugz/mdma.html
[7] The Complete Book of Ecstacy: Synthesis Books; P.O.Box 610341; Birmingham, AL 35261, USA, (1992)
[8] 1) Saunders.N.: E for Ecstasy, Published by Nicholas Saunders, 14 Neal's Yard, London, WC2H 9DP, UK, (1993)
2) Eisner, B.: Ecstasy-MDMA story, Ronin Publishing, Inc., Box 1035, Berkeley CA, 94701, (1994)
3) Saunders, N.: Ecstasy and the Dance Culture, Published by Nicholas Saunders, 14 Neal's Yard, London, WC2H 9DP, UK, (1995)
4) Saunders, N.: Ecstasy Reconsidered, Turnaround, London N22 6TZ, UK, (1997)
[9] http:// hyperreal.com/drugs/e4x/

2.1.1.2 Data Sheets: MDMA and its Derivatives

Substance Number 1 in Chapter 2.1.1

IUPAC Name: 2-Benzo[1,3]dioxol-5-yl-1-methyl-ethylamine
Synonyms: 1-(3,4-Methylenedioxyphenyl)-2-propanamine
3,4-Methylenedioxyamphetamine
1-Amino-2-(3,4-methylenedioxyphenyl)propane
α-Methyl-3,4-methylenedioxyphenethylamine
3,4-Methylenedioxyphenylisopropylamine
Tenamfetamine

Chemical Structure:

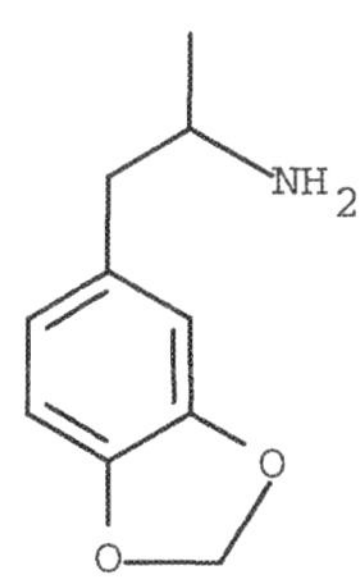

MF: C10H13NO2
MW: 179.218
CA Registry Number:[4764-17-4]; racemate [51497-09-7]; [R]-[61614-60-6]; [S]-[65620-66-8]; hydrochloride racemate [6292-91-73]
CA Chemical Name: 1,3-Benzodioxole-5-ethanamine, α-methyl-
Category: Psychotomimetic phenethylamine
Street Names: MDA; Zen; Love drug
Abuse: Frequent
Type of action: Psychotomimetic, hallucinogen, entactogen, euphoriant
Human active dose: 80-150 mg
Duration of action: 8-10 hours
Toxic manifestations: Hyperthermia, anxiety, confusion, dehydration, convulsions, cardiovascular disturbances, neurotoxicity
Toxicity: LD50 is 65 mg/kg i.p. (mouse) [1].

This psychotomimetic amphetamine shows pronounced entactogenic properties. Being easily available from several chemical suppliers, this drug was very popular in the USA in the 1970s. Similar to MDMA, it was used extensively in psychotherapy [2]. Today, this substance appears occasionally as a substitute for MDMA [2]. More particularly, it is employed in some religious groups to improve the ability to meditate [3]. The identification of this drug by GC-MS and HPLC [4] has been described.

References:

[1] Ho, T. Beng, et al.: J. Med. Chem., **13**, p.26, (1970)
[2] McCann, Una; Ricaurte, G.A.: Use and Abuse of Ring Substituted Amphetamines. In: Amphetamine and its analogs; Cho, A.K. and Segal, D.S. (editors); Academic Press, p.378, (1994)
[3] Watson, L.; Beck, J.: J. of Psychoactive Drugs, **23** (3), p.261, (1991) and http://www.ecstasy. org/god.html
[4] Mills III, T.; Roberson, J.C.; McCurdy, H.H.; Wall, W.H.: Instrumental Data for Drug Analysis, Elsevier Science Publishing Co., Inc., p.1460, (1992)

Substance Number 2 in Chapter 2.1.1

IUPAC Name: (2-Benzo[1,3]dioxol-5-yl-1-methyl-ethyl)-methyl-amine
Synonyms: N-Methyl-1-(3,4-methylenedioxyphenyl)-2-propanamine
3,4-Methylenedioxymethamphetamine
1-(Methylamino)-2-(3,4,-methylenedioxyphenyl)propane
N, α-Dimethyl-3,4-methylenedioxyphenethylamine
N-Methyl-3,4-methylenedioxyphenylisopropylamine

Chemical structure:

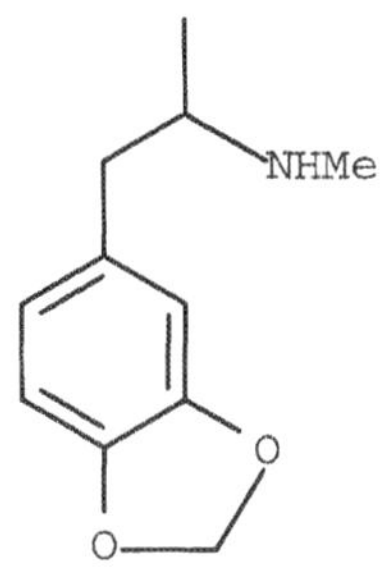

MF: C11H15NO2
MW: 193.245
CA Registry Number: [42542-10-9]; racemate [69610-10-2]; [R]-[81262-70-6]; [S]-[66142-89-0]; hydrochloride [64057-70-1]
CA Chemical Name: 1,3-Benzodioxole-5-ethanamine, N,α-dimethyl-
Category: Psychotomimetic phenethylamine
Street Names: MDMA; Adam; MD; Ecstasy; XTC; X; A Bean; Baby slits; Chocolate chips; Clarity; Deccadence; Doctor; E; Elaine; Essence; Euphoria; Eve; Roll; Rolling; Running; Slits; Speed for lovers
Abuse: Very frequent
Type of action: Entactogen, euphoriant, psychotomimetic
Human active dose: 80-200 mg
Duration of action: 4-6 hours
Toxic manifestations: Hyperthermia, anxiety, confusion, dehydration, marked anorexia, neurotoxicity
Toxicity: LD50 is 97 mg/kg i.p. (mouse); 49 mg/kg i.p. (rat); 14 mg/kg i.v. (dog); 22 mg/kg i.v. (monkey) [1].

This is a well known world wide abused drug, mainly due to its ability to produce strong euphoria and to improve communication and contacts (« entactogen ») [2] among participants at the « Rave parties ». The security range of this drug seems to be particularly narrow. Serious disturbances of cardiovascular function may occur at doses as low as 300 mg, especially when considerable physical effort in a hot environment is involved [3]. There is a huge amount of information on this substance in the official literature [4,7] and on the Internet [5] as well.

Recently, sophisticated methods of administration of MDMA were described [6]. Thus, a dose of MDMA is followed by another lower dose of MDMA (« booster »; typically 120 and 50 mg) or by another psychotomimetic drug such as 2CB, LSD, THC [8]. In the last case, new, unpredictable and potentially dangerous effects are usually experienced. Apart from its abuse in dancing clubs, this drug is employed, usually at low dose, to promote meditation in some religious groups, particularly of the Zen inspiration [9,10]. The identification of this drug by GC [11], GC-MS and HPLC [12] has been thoroughly described.

References:

[1] Merck Index, 11th Edition; Merck & Co. Inc., N.J., p. 904, (item 5646), (1989)
[2] Nichols, D.E; Hoffman, A.J.; Oberlender, R.A.; Jacob, Peyton III; Shulgin, A.T.: J. Med. Chem., **29**, p.2009, (1986)
[3] Hayner, G.H.; McKinney, H.: J. of Psychoactive Drugs, **18**, p.341, (1986)
[4] Journal of Psychoactive Drugs, **18**, No. 4, (1986)
[5] http://www.ecstasy.org
[6] Shulgin, A.T. and Shulgin, A.: PIHKAL, Transform Press, Inc., P.O.Box 13675, Berkeley, CA, p.737, (1991)
[7] Schmid, J.C.: Neurochemistry of Ring substituted Amphetamine Analogs. In: Amphetamine and its analogs; Cho, A.K. and Segal, D.S. (editors), Academic Press, pp.151-171, (Chapter 5), (1994)
[8] http:// www.ecstasy org/combinations.html
[9] Watson, L.; Beck, J.: J. of Psychoactive Drugs, **23**, p.261, (1991)
[10] http:// www.ecstasy.org/god.html
[11] Bost, R.O.: J. Forensic Sci., **33**, p.576, (1988)
[12] Noggle, F.T.; Clark. C.R.; Valaer, A.K.; DeRuiter, J.: J. Chromatogr. Sci., **26**, p.410, (1988)

Substance Number 3 in Chapter 2.1.1

IUPAC Name: (2-Benzo[1,3]dioxol-5-yl-1-methyl-ethyl)-ethyl-amine
Synonyms: N-Ethyl-1-(3,4-methylenedioxyphenyl)-2-propanamine
N-Ethyl-3,4-methylenedioxyamphetamine
3,4 -Methylenedioxyethamphetamine
1-(Ethylamino)-2-(3,4-methylenedioxyphenyl)propane
N-Ethyl- α-methyl-3,4-methylenedioxyphenethylamine
N-Ethyl-3,4-methylenedioxyphenylisopropylamine

Chemical structure:

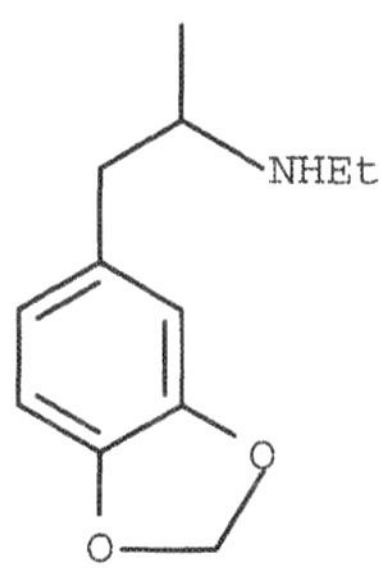

MF: $C_{12}H_{17}NO_2$
MW: 207.272
CA Registry Number:[14089-52-2]; racemate [82801-81-8]; [R]-[114612-27-0]; [S]-[114612-26-9]; hydrochloride [74341-78-9]
CA Chemical Name: 1,3-Benzodioxole-5-ethanamine, N-ethyl-α-methyl-
Category: Psychotomimetic phenethylamine
Street Names: MDEA; Eve
Abuse: Frequent
Type of action: Entactogen, euphoriant, psychotomimetic
Human active dose: 100-200 mg
Duration of action: 3-5 hours
Toxic manifestations: Hyperthermia, anxiety, anorexia, confusion, dehydration, convulsions, cardiovascular disturbances, neurotoxicity
Toxicity: Not reported

This compound seems to have lower incidence of undesirable effects and be less neurotoxic than MDMA [1]. This MDMA analogue has been created to circumvent the use of methylamine which is a highly suspect and controlled precursor for several drugs of abuse. The substance is one of the most suitable substitutes of MDMA because of their practically identical pharmacological profiles [2]. Only, MDEA seems to be a weaker CNS stimulant than MDMA [3]. The last reference also describes the synthesis of MDEA and several of its analogues. The identification of this drug by GC-MS and HPLC [4] has been described.

References:

[1] Ricaurte, G.A.; Finnegan, K.F.; Nichols, D.E.; DeLanney, L.E.; Irwin, I.; Langston, J.W.: Eur. J. Pharmacol., **137**, p.265, (1987)
[2] McCann, Una; Ricaurte, G.A.: Use and Abuse of Ring-Substituted Amphetamines. In: Amphetamine and its analogs; Cho, A.K. and Segal, D.S. (editors); Academic Press, p.379, (Chapter 12), (1994)
[3] Braun, U.; Shulgin, A.T.; Braun, G.: J. Pharm. Sci., **69**, p.192, (1980)
[4] Mills III, T.; Roberson, J.C.; McCurdy, H.H.; Wall, W.H.: Instrumental Data for Drug Analysis, Elsevier Science Publishing Co., Inc., p.1462, (1992)

Substance Number 4 in Chapter 2.1.1

IUPAC Name: N-(2-Benzo[1,3]dioxol-5-yl-1-methyl-ethyl)-hydroxylamine
Synonyms: N-Hydroxy-1-(3,4-methylenedioxyphenyl)-2-propanamine
N-Hydroxy-MDA
N-Hydroxy-3,4-methylenedioxyamphetamine
1-(Hydroxyamino)-2-(3,4-methylenedioxyphenyl)propane
N-Hydroxy-α-methyl-3,4-methylenedioxyphenethylamine
N-Hydroxy-3,4-methylenedioxyphenylisopropylamine
N-(α-Methyl-3,4-methylenedioxyphenethyl)hydroxylamine

Chemical structure:

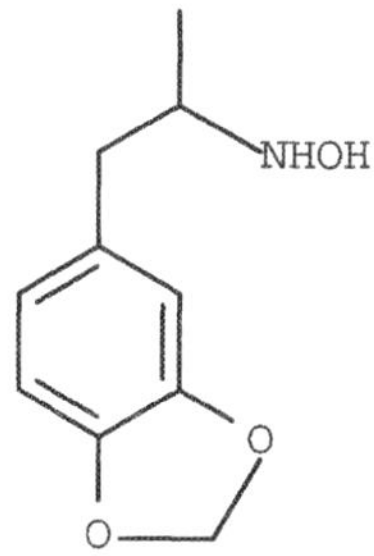

MF: $C_{10}H_{13}NO_3$
MW: 195.217
CA Registry Number: [74698 47 8]; racemate [114562-59-3]
CA Chemical Name: 1,3-Benzodioxole-5-ethanamine, N-hydroxy-α-methyl-
Category: Psychotomimetic phenethylamine
Street Names: MDAOH
Abuse: Frequent
Type of action: Psychotomimetic, entactogen, euphoriant
Human active dose: 80-200 mg
Duration of action: 6-8 hours
Toxic manifestations: Cardiovascular disturbances, hyperthermia, anxiety, confusion, dehydration, convulsions, neurotoxicity
Toxicity: Not reported

The pharmacological profile of this substance closely resembles that of MDA [1]. Being particularly easy to synthesise, it serves frequently as an MDMA or MDA substitute [2,3]. The identification of this drug by MS and HPLC has been described [2,4].

References:

[1] Braun, U.; Shulgin, A.T.; Braun, G.: J. Pharm. Sci., **69**, p.192, (1980)
[2] Noggle, F.T.; Clark, C.R.; Valaer, A.K.; DeRuiter, J.: J. Chromatogr. Sci., **26**, p.410, (1988)

[3] Valaer, A.K.; Ravis, W.R.; Clark, C.R.: J. Chromatogr. Sci., **28**, p.482, (1990)
[4] Mills III, T.; Roberson, J.C.; McCurdy, H.H.; Wall, W.H.: Instrumental Data for Drug Analysis, Elsevier Science Publishing Co., Inc., p.226, (1992)

Substance Number 5 in Chapter 2.1.1

IUPAC Name: (2-Benzo[1,3]dioxol-5-yl-1-methyl-ethyl)-propylamine
Synonyms: 1-(3,4-Methylenedioxyphenyl)-N-propyl-2-propanamine
3,4-Methylenedioxy-N-propylamphetamine
1-(3,4-Methylenedioxyphenyl)-2-propylaminopropane
α-Methyl-3,4-methylenedioxy-N-propylphenethylamine
3,4-Methylenedioxyphenyl-N-propylisopropylamine

Chemical structure:

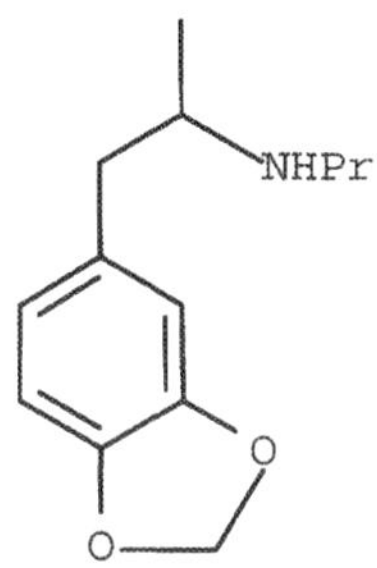

MF: C13H19NO2
MW: 221.299
CA Registry Number: [74698-36-5]
CA Chemical Name: 1,3-Benzodioxole-5-ethanamine, α-methyl-N-propyl-
Category: Psychotomimetic phenethylamine
Street Names: MDPR
Abuse: Uncommon
Type of action: « Primer » for various psychotomimetics
Human active dose: 150-200 mg
Duration of action: Unknown
Toxic manifestations: Cardiovascular disturbances, neurotoxicity
Toxicity: Not reported

A seizure of this substance has been reported [1]. Despite being a close MDMA analogue, it has no particular action in man [2] but it strongly enhances the action of LSD [3]. The identification of this drug by MS and HPLC has been described [4].

References:

[1] Microgram, XXI, (8), p.114, (1988)
[2] Braun, U.; Shulgin, A.T.; Braun, G.: J. Pharm. Sci., **69**, p.192, (1980)
[3] Shulgin, A.T. and Shulgin, A.: PIHKAL, Transform Press, Inc., P.O.Box 13675, Berkeley, CA, p.754, (1991)
[4] Mills III, T.; Roberson, J.C.; McCurdy, H.H.; Wall, W.H.: Instrumental Data for Drug Analysis, Elsevier Science Publishing Co., Inc., p.270, (1992)

2.1.2 Substitutes of MDMA

This chapter is devoted to further designer drugs possessing the MDMA-like effect in man. Recently, in research for entactogenic substances of low neurotoxicity [1], several new MDA homologues were synthesised. Apart from psychic activity, these compounds seem to show similar physiological effects (cardiovascular disturbances, hyperthermia and dehydratation) to MDMA and its derivatives but of lower intensity.

Some of them have appeared on the clandestine market. This is particularly the case for substances such as MDBA (**1**, 1-(3,4-methylenedioxyphenyl)-2-aminobutane), its N-methyl analogue, MBDB, « EDEN » (**2**, N-methyl-1-(3,4-methylenedioxyphenyl)-2-aminobutane), and the N-hydroxy derivative of MDMA (**3**) [2,3].

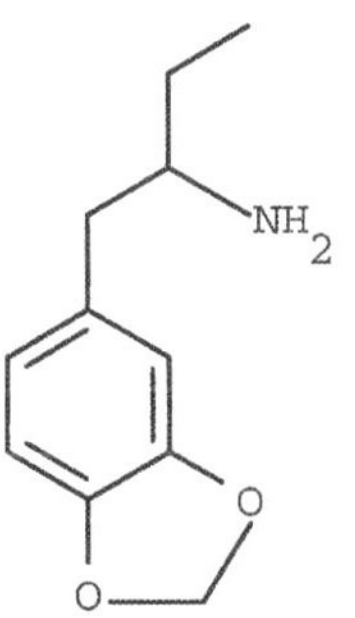

1

150-230 mg

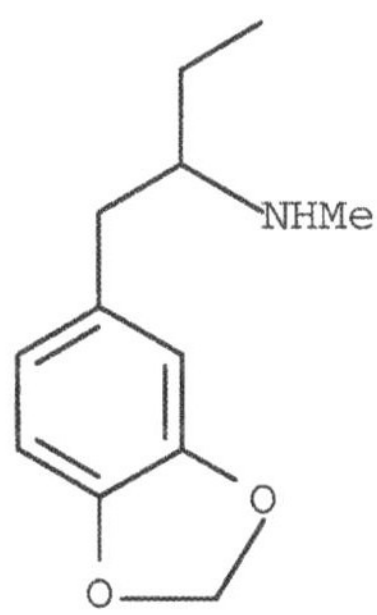

2

130-280 mg

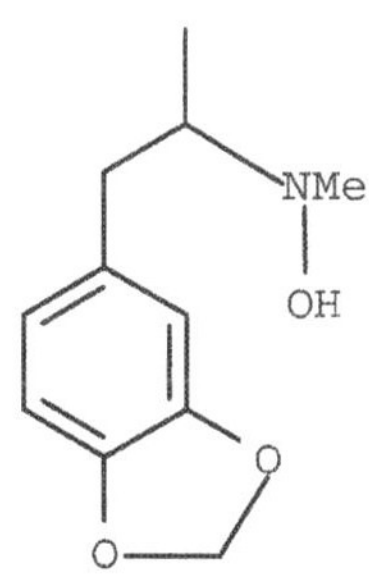

3

100-160 mg

Two other compounds, which actually are not phenethylamines but benzylamines, are also potent entactogens. Interestingly, they do not show the powerful anorexic action typical for MDA derivatives. These substances have also appeared on the drug scene (ALPHA, **4** and MALPHA, **5**) [2,4].

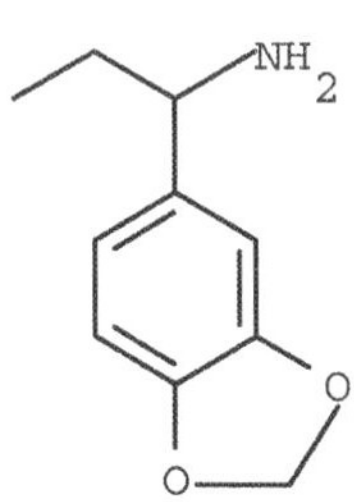

4

100-150 mg

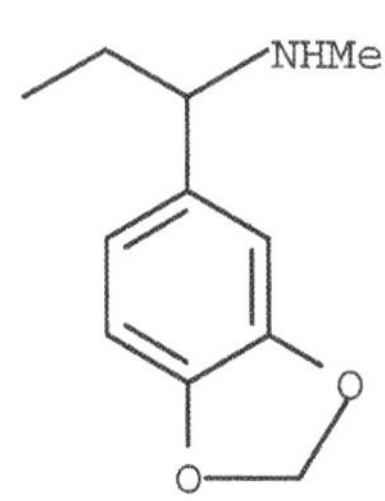

5

60-100 mg

Examination of the compounds above might lead to a conclusion that the entactogenic action is closely linked and limited to 1,3-benzodioxole derivatives.

Nevertheless, this particular activity is more or less shown by several other phenethylamines.

Thus, the following phenethylamines (**6**), 4-ethoxy-3-methoxyphenethylamine; (**7**), 4-ethoxy-3-methoxyamphetamine; (**8**), 3,4,5-trimethoxyamphetamine; (**9**), 2,5-dimethoxy-4-nitrophenethylamine; (**10**), 2,5-dimethoxy-4-methylthiophenethylamine, and (**11**), 4-allyloxy-3,5-dimethoxyphenethylamine have been reported to have more or less entactogenic properties and produce marked euphoria [5,6]. Some of these drugs are sometimes present on the « Rave » scene.

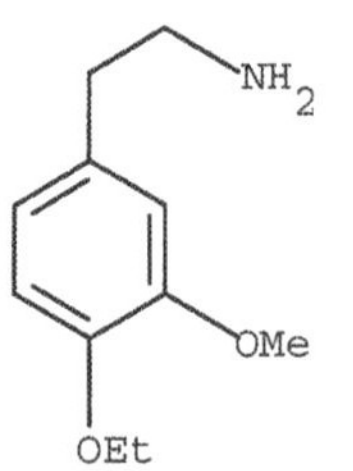

6

100-300 mg

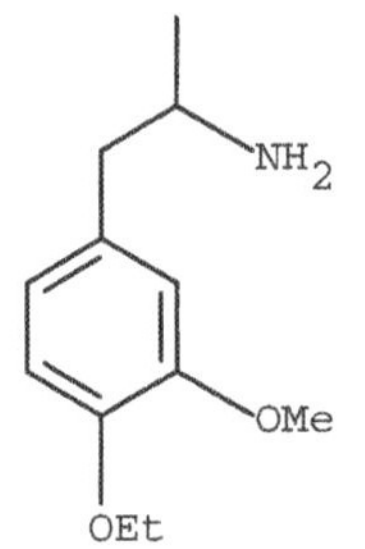

7

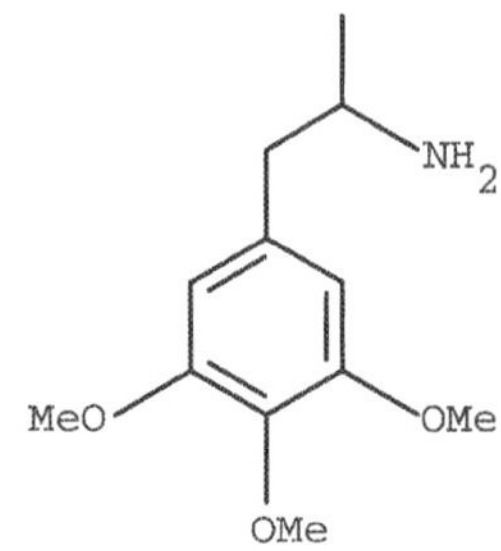

8

120-200 mg

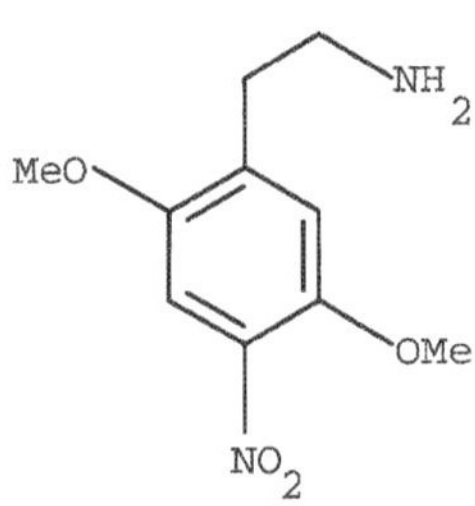

9

100-150 mg

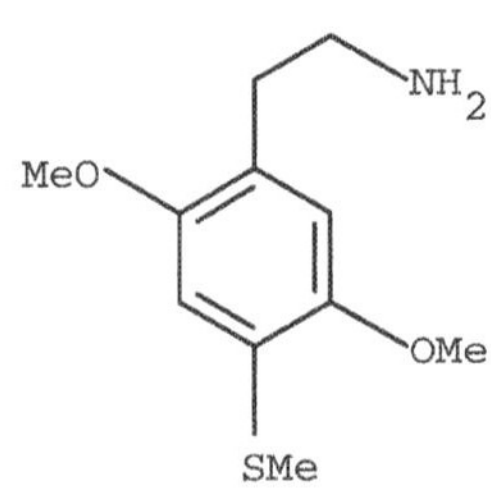

10

60-100 mg

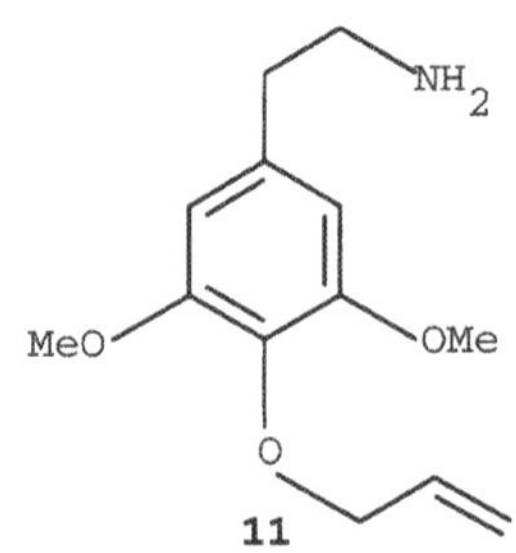

11

20-35 mg

Also, 2,5-dimethoxyamphetamine, 2,4,5-trimethoxyamphetamine (described in the chapter about other psychotomimetic phenethylamines) and at least one tryptamine (etryptamine) have been reported appearing in similar circumstances. This last compound is described in the chapter on psychotomimetic indolealkylamines.

2.1.2.1 General References

[1] Nichols, D.E.; Hoffman, A.J.; Oberlender, R.A.; Jacob, Peyton III; Shulgin, A.T.: J. Med. Chem., **29**, p.2009, (1986)
[2] DeRuiter, J.; Clark, C.R.; Noggle, F.T.: J. Chromatogr. Sci., **29**, p.103, (1990)
[3] Shulgin, A.T. and Shulgin, A.: PIHKAL, Transform Press, Inc., P.O.Box 13675, Berkeley, CA, p.698 and 778, (1991)
[4] King, L.A. et al.: Forensic Sci. Int., **77**, p.141, (1996)
[5] Leminger, O.: Chem. Pr., **22**, p.553, (1972)
[6] Ref. [3], the notes on these compounds in the Book II

2.1.2.2 Data Sheets: Substitutes of MDMA

Substance Number 1 in Chapter 2.1.2

IUPAC Name: 1-Benzo[1,3]dioxol-5-ylmethyl-propylamine
Synonyms: 1-(3,4-Methylenedioxyphenyl)-2-butanamine
α-Ethyl-3,4-methylenedioxyphenethylamine
2-Amino-1-(3,4-methylenedioxyphenyl)butane
1-(3,4-Methylenedioxyphenyl)-2-aminobutane

Chemical structure:

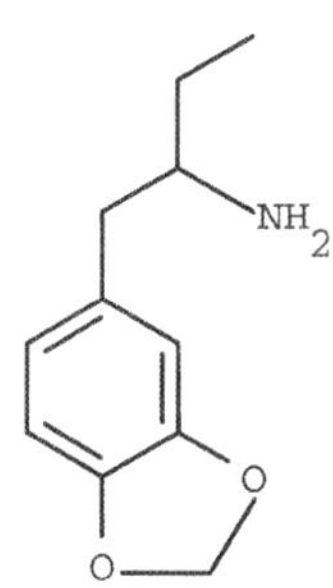

MF: C11H15NO2
MW: 193.245
CA Registry Number: [107447-03-0]; racemate [103818-45-7]; [R]-[103882-51-5]; [S]-[103882-52-6]; hydrochloride racemate [103818-36-6]
CA Chemical Name: 1,3-Benzodioxole-5-ethanamine, α-ethyl-
Category: Psychotomimetic phenethylamine
Street Names: J; BDB; MDBA
Abuse: Frequent
Type of action: Entactogen, euphoriant, psychotomimetic
Human active dose: 150-230 mg
Duration of action: 3-5 hours
Toxic manifestations: Cardiovascular disturbances, hyperthermia, anxiety, confusion, dehydration, convulsions
Toxicity: Not reported

This drug and its N-methyl derivative, MBDB, often replace MDMA. These compounds seem to be less neurotoxic than MDMA [1]. This homologue of MDMA appeared on the US clandestine drug market in 1986 [2]. This reference also describes the identification of this drug by MS and HPLC. Having practically the same effect in man as MDMA [3], this substance is one of its most suitable substitutes. However, MDBA seems to be a weaker CNS stimulant than MDMA [3].

References:

[1] Nichols, D.E; Hoffman, A.J.; Oberlender, R.A.; Jacob, Peyton III; Shulgin, A.T.: J. Med. Chem., **29**, p.2009, (1986)
[2] DeRuiter, J.; Clark, C.R.; Noggle, F.T.: J. Chromatogr., Sci. **29**, p.103, (1991)
[3] Shulgin, A.T. and Shulgin, A.: PIHKAL, Transform Press, Inc., P.O.Box 13675, Berkeley, CA, p.698, (1991)

Substance Number 2 in Chapter 2.1.2

IUPAC Name: (1-Benzo[1,3]dioxol-5-ylmethyl-propyl)-methyl-amine
Synonyms: N-Methyl-1-(3,4-methylenedioxyphenyl)-2-butanamine
2-(Methylamino)-1-(3,4-methylenedioxyphenyl)butane
α-Ethyl-N-methyl-3,4-methylenedioxyphenethylamine
N-Methyl-1-(3,4-methylenedioxyphenyl)-2-aminobutane

Chemical structure:

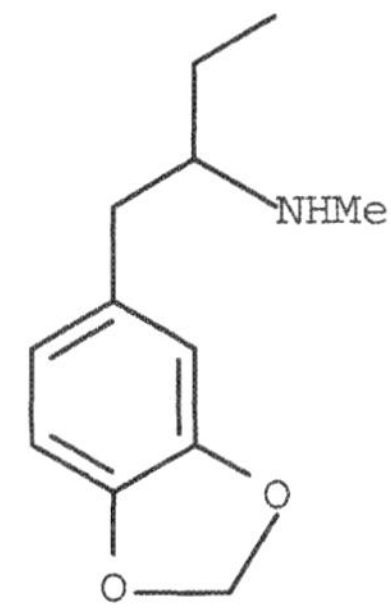

MF: C12H17NO2
MW: 207.272
CA Registry Number: [135795-90-3], racemate [103818-46-8]; hydrochloride [R]-[103882-49-1]; hydrochloride [S]-103882-50-4]; hydrochloride, racemate [103818-37-7]
CA Chemical Name: 1,3-Benzodioxole-5-ethanamine, N-methyl-α-ethyl-
Category: Psychotomimetic phenethylamine
Street Names: Eden; MJ; MBDB; Methyl-J
Abuse: Frequent
Type of action: Entactogen, euphoriant, psychotomimetic
Human active dose: 130-280 mg
Duration of action: 4-6 hours
Toxic manifestations: Hyperthermia, anxiety, confusion, dehydration, convulsions, cardiovascular disturbances
Toxicity: LD50 is 120 mg/kg i.p. (mouse).

This drug and its N-desmethyl derivative, MDBA, often replace MDMA. These compounds seem to be less neurotoxic than MDMA [1]. This homologue of MDMA appeared on the US clandestine drug market in 1986 [2]. This reference also describes the identification of this drug by MS and HPLC. In 1995, the drug was detected in Switzerland. This substance is one of the most suitable substitutes for MDMA because of their nearly identical pharmacological properties, but it seems to be a weaker CNS stimulant [3].

References:

[1] Nichols, D.E; Hoffman, A.J.; Oberlender, R.A.; Jacob, Peyton III; Shulgin, A.T.: J. Med. Chem., **29**, p.2009, (1986)

[2] DeRuiter, J.; Clark, C.R.; Noggle, F.T.: J. Chromatogr. Sci., **29**, p.103, (1990)

[3] Shulgin, A.T. and Shulgin, A.: PIHKAL, Transform Press, Inc., P.O.Box 13675, Berkeley, CA, p.778, (1991)

Substance Number 3 in Chapter 2.1.2

IUPAC Name: N-(2-Benzo[1,3]dioxol-5-yl-1-methyl-ethyl)-N-methyl-hydroxylamine
Synonyms: N-Hydroxy-N-methyl-1-(3,4-methylenedioxyphenyl)-2-propanamine
N-hydroxy-MDMA
N-Hydroxy-3,4-methylenedioxymethamphetamine
N-Hydroxy-N,α-dimethyl-3,4-methylenedioxyphenethylamine
N-Hydroxy-N-methyl-3,4-methylenedioxyphenylisopropylamine
N-Methyl-N-(α-methyl-3,4-methylenedioxyphenethyl)hydroxylamine

Chemical structure:

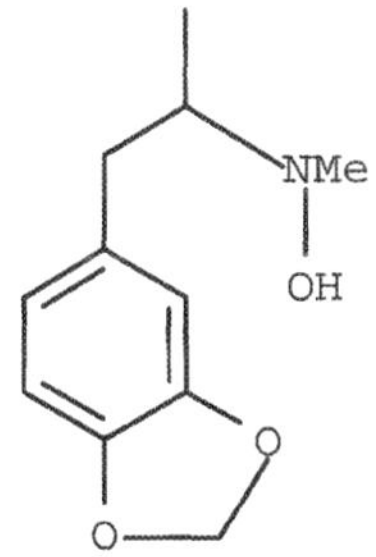

MF: C11H15NO3
MW: 209.244
CA Registry Number: [74698-47-8]; hydrochloride [74341-83-6]
CA Chemical Name: 1,3-Benzodioxole-5-ethanamine, N-hydroxy-N,α-dimethyl-
Category: Psychotomimetic phenethylamine
Street Names: FLEA; MDMAOH
Abuse: Rare
Type of action: Entactogen, euphoriant, psychotomimetic
Human active dose: 100-160 mg
Duration of action: 3-5 hours
Toxic manifestations: Hyperthermia, anxiety, confusion, dehydration, convulsions, cardiovascular disturbances, neurotoxicity
Toxicity: Not reported

This substance shows MDMA-like action in man [1]. Being very easy to prepare, this drug is one of the most suitable MDMA substitutes [3]. Attempts of large scale clandestine production in Holland have been reported [2,3]. The last reference also describes the identification of this drug by MS and HPLC.

References:

[1] Shulgin, A.T. and Shulgin, A.: PIHKAL, Transform Press, Inc., P.O.Box 13675, Berkeley, CA, p.671, (1991)
[2] Verweij, A.M.A.: Microgram, XXVIII, (7), p.224, (1995)
[3] Noggle, F.T.; Clark, C.R.; DeRuiter, J.; Cain, P.: Microgram, XXIX, (1), p.10, (1996)

Substance Number 4 in Chapter 2.1.2

IUPAC Name: 1-Benzo[1,3]dioxol-5-yl-propylamine
Synonyms: 1-(3,4-Methylenedioxyphenyl)-1-propanamine
α-Ethyl-(3,4-methylenedioxy)benzylamine
1-Amino-1-(3,4-methylenedioxyphenyl)propane

Chemical structure:

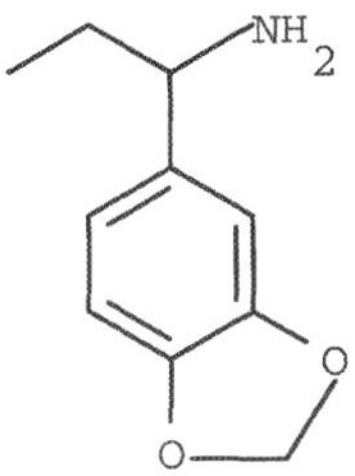

MF: C10H13NO2
MW: 179.218
CA Registry Number: [127292-42-6]
CA Chemical Name: 1,3-Benzodioxole-5-methanamine-α-ethyl-
Category: Psychotomimetic phenethylamine
Street Names: ALPHA
Abuse: Limited
Type of action: Entactogen, euphoriant, psychotomimetic
Human active dose: 100-150 mg
Duration of action: 2-4 hours
Toxic manifestations: Not reported
Toxicity: Not reported

Interestingly, this benzylamine possesses MDMA-like effects in man but lacks its powerful anorectic activity [1]. It may be easily prepared from readily available and uncontrolled precursors [2]. This reference also describes the synthesis and identification of this drug by MS and HPLC. A seizure of this substance by Dutch authorities has been reported [3].

References:

[1] Shulgin, A.T. and Shulgin, A.: PIHKAL, Transform Press, Inc., P.O.Box 13675, Berkeley, CA, p.718, (1991)
[2] DeRuiter, J.; Clark, C.R.; Noggle, F.T.: J. Chromatogr. Sci., **28**, p.129, (1990)
[3] King, L.A. et al.: Forensic Sci. Int., **77**, p.141, (1996)

Substance Number 5 in Chapter 2.1.2

IUPAC Name: (1-Benzo[1,3]dioxol-5-yl-propyl)-methyl-amine
Synonyms: N-Methyl-1-(3,4-methylenedioxyphenyl)-1-propanamine
Ethyl-α-N-methyl-3,4-methylenedioxybenzylamine
1-(Methylamino)-1-(3,4-methylenedioxyphenyl)propane

Chemical structure:

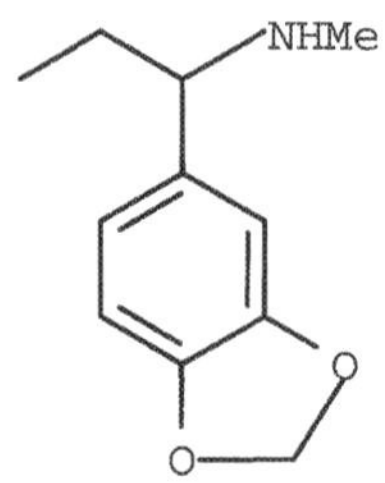

MF: C11H15NO2
MW: 193.245
CA Registry Number: [127292-43-7]
CA Chemical Name: 1,3-Benzodioxole-5-methanamine, α-ethyl-N-methyl-
Category: Psychotomimetic phenethylamine
Street Names: MALPHA
Abuse: Limited
Type of action: Entactogen, euphoriant, psychotomimetic
Human active dose: 60-100 mg
Duration of action: 4-6 hours
Toxic manifestations: Not reported
Toxicity: Not reported

Similar to its N-desmethyl analogue, ALPHA, described above, this benzylamine also possesses MDMA-like effects in man and lacks its powerful anorectic activity [1]. It may be easily prepared from readily available precursors [2]. This reference also describes the identification of this drug by MS and HPLC.

References:

[1] Shulgin, A.T. and Shulgin, A.: PIHKAL, Transform Press, Inc., P.O.Box 13675, Berkeley, CA, p.718, (1991)
[2] DeRuiter, J.; Clark, C.R.; Noggle, F.T.: J. Chromatogr. Sci., **28**, p.129, (1990)

Substance Number 6 in Chapter 2.1.2

IUPAC Name: 2-(4-Ethoxy-3-methoxy-phenyl)-ethylamine
Synonyms: 2-(4-Ethoxy-3-methoxyphenyl)ethanamine
4-Ethoxy-3-methoxyphenethylamine

Chemical structure:

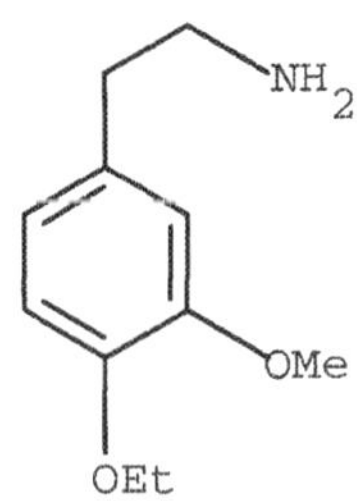

MF: C11H17NO2
MW: 195.261
CA Registry Number: [36377-59-0]
CA Chemical Name: Benzeneethanamine, 4-ethoxy-3-methoxy-
Category: Psychotomimetic phenethylamine
Street Names: MEPEA
Abuse: Rare
Type of action: Euphoriant, psychotomimetic
Human active dose: 100-300 mg
Duration of action: 2-3 hours
Toxic manifestations: Not reported
Toxicity: Not reported

This drug used as a « mood enhancer » is an euphoriant of low toxicity [1]. It may be easily be prepared from a widely available precursor (vanilline). The IR spectrum of this substance has been described [1].

References:

[1] Leminger, O.: Chem. Pr., **22**, p.553, (1972)
[2] Shulgin, A.T. and Shulgin, A.: PIHKAL, Transform Press, Inc., P.O.Box 13675, Berkeley, CA, p.769, (1991)

Substance Number 7 in Chapter 2.1.2

IUPAC Name: 2-(4-Ethoxy-3-methoxy-phenyl)-1-methyl-ethylamine
Synonyms: 1-(4-Ethoxy-3-methoxyphenyl)-2-propanamine
4-Ethoxy-3-methoxyamphetamine
α-Methyl-4-ethoxy-3-methoxyphenethylamine
4-Ethoxy-3-methoxyphenylisopropylamine
1-Amino-2-(4-ethoxy-3-methoxyphenyl)propane

Chemical structure:

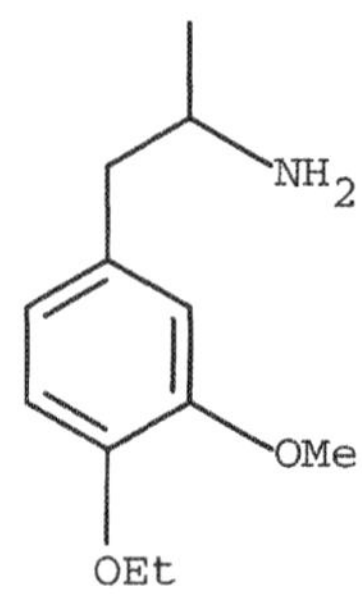

MF: C12H19NO2
MW: 209.288
CA Registry Number: [32560-70-6]; racemate [129693-62-5]
CA Chemical Name: Benzeneethanamine, 4-ethoxy-3-methoxy-α-methyl-
Category: Psychotomimetic phenethylamine
Street Names: 4EMA
Abuse: Rare
Type of action: Euphoriant, psychotomimetic
Human active dose: Not reported
Duration of action: Not reported
Toxic manifestations: Cardiovascular disturbances, nausea, hyperthermia, anxiety, confusion, dehydration
Toxicity: Not reported

At present, there is practically no evidence for abuse of this amphetamine which is a known compound [1,2]. However, abuse is very likely to become more frequent in the near future because of the easy availability of this amphetamine. It can be easily prepared from commercially available precursors [2] (vanillin or eugenol). Moreover, two other easily available and closely related 4-ethoxyamphetamines (4-ethoxyamphetamine and 4-ethoxy-2,5-dimethoxy-amphetamine) have already become well known drugs of abuse.

References:

[1] Shepard, E.R.; Noth, J.F.; Herschel, D.P.; Simmans,K.C.: J. Am. Chem. Soc., **74**, p.4611, (1952)
[2] Mekenyan, O.; Mercier, C.; Bonchev, D.; Dubois, J. E.: Eur. J. Med. Chem., **28**, p.811, (1993)
[3] Shulgin, A.T. and Shulgin, A.: PIHKAL, Transform Press, Inc., P.O.Box 13675, Berkeley, CA, p.769, (1991)

Substance Number 8 in Chapter 2.1.2

IUPAC Name: 1-Methyl-2-(3,4,5-trimethoxy-phenyl)-ethylamine
Synonyms: (3,4,5-Trimethoxyphenyl)-2-propanamine
3,4,5-Trimethoxyamphetamine
1-Amino-2-(3,4,5-trimethoxyphenyl)propane
α-Methyl-3,4,5-trimethoxyphenethylamine
3,4,5-Trimethoxyphenylisopropylamine

Chemical structure:

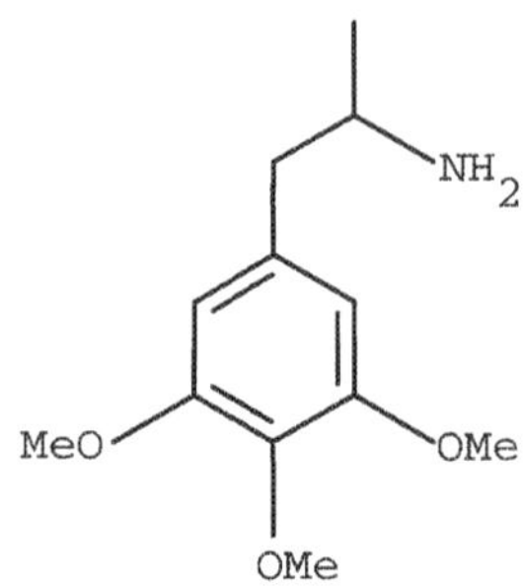

MF: C12H19NO3
MW: 225.287
CA Registry Number: [1082-88-8]; racemate [22199-17-3]
CA Chemical Name: Benzeneethanamine, 3,4,5,trimethoxy-α-methyl-
Category: Psychotomimetic phenethylamine
Street Names: TMA
Abuse: Limited
Type of action: Psychotomimetic, euphoriant
Human active dose: 120-200 mg
Duration of action: 6-8 hours
Toxic manifestations: Anxiety, confusion, violence, cardiovascular disturbances, nausea, hyperthermia, dehydration
Toxicity: LD50 is 250 mg i.p. (mouse) [1].

In man, this amphetamine is a powerful euphoriant but it may also produce attacks of violence [2]. This drug has recently appeared in Germany in connection with the "Rave" subculture. Its abuse may become more frequent in the future because its key precursor, 3,4,5,-trimethoxybenzaldehyde, has become readily available. The identification of this drug by MS and HPLC has been described [3].

References:

[1] Ho, T. Beng, et al.: J. Med. Chem., **13**, p.26, (1970)
[2] Shulgin, A.T.: Psychotomimetic drugs: Structure-Activity Relationships. In: Handbook of Psychopharmacology, Plenum Press, New York, Vol.**11**, p.283, (1978)
[3] Mills III, T.; Roberson, J.C.; McCurdy, H.H.; Wall, W.H.: Instrumental Data for Drug Analysis, Elsevier Science Publishing Co., Inc., p.2326, (1992)

Substance Number 9 in Chapter 2.1.2

IUPAC Name: 2-(2,5-Dimethoxy-4-nitro-phenyl)-ethylamine
Synonyms: 2-(2,5-Dimethoxy-4-nitrophenyl)ethanamine
2,5-Dimethoxy-4-nitrophenethylamine

Chemical structure:

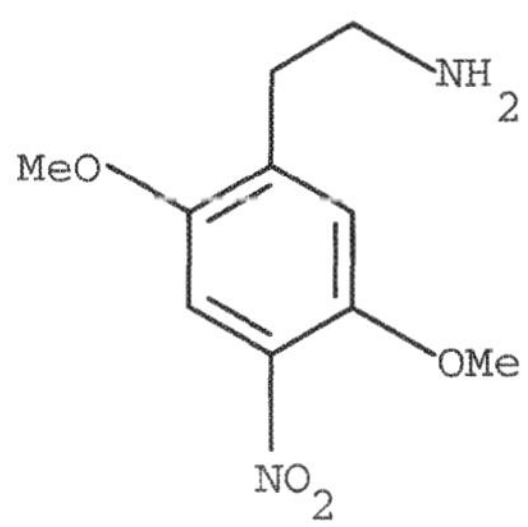

MF: C10H14N2O4
MW: 226.232
CA Registry Number: Not reported
CA Chemical Name: Benzeneethanamine, 2,5-dimethoxy-4-nitro-
Category: Psychotomimetic phenethylamine
Street Names: 2CN
Abuse: Rare
Type of action: Euphoriant, psychotomimetic, entactogen
Human active dose: 100-150 mg
Duration of action: 4-6 hours
Toxic manifestations: Cardiovascular disturbances, nausea, dehydration
Toxicity: Not reported

This rarely abused substance has been reported to produce MDMA-like action in man [1]. It can be prepared from the same precursor as 4-bromo-2,5-dimethoxyphenethylamine (2CB).

References:

[1] Shulgin, A.T. and Shulgin, A.: PIHKAL, Transform Press, Inc., P.O.Box 13675, Berkeley CA, p.542, (1991)

Substance Number 10 in Chapter 2.1.2

IUPAC Name: 2-(2,5-Dimethoxy-4-methylsulfanyl-phenyl)-ethylamine
Synonyms: 2-(2,5-Dimethoxy-4-methylthiophenyl)ethanamine
2,5-Dimethoxy-4-methylthiophenethylamine

Chemical structure:

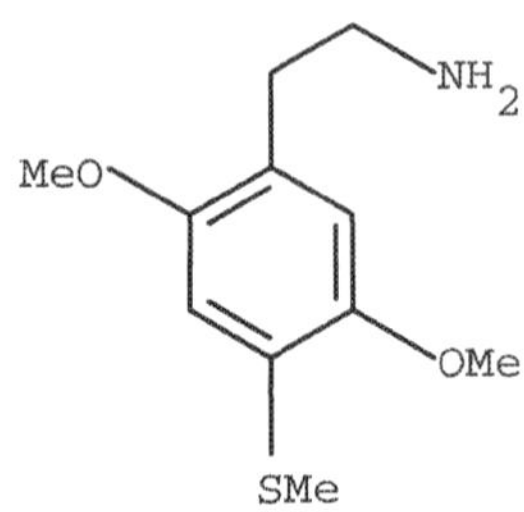

MF: C11H17NO2S
MW: 227.321
CA Registry Number: [61638-09-3]
CA Chemical Name: Benzeneethanamine, 2,5-dimethoxy-4-methylthio-
Category: Psychotomimetic phenethylamine
Street Names: 2CT
Abuse: Rare
Type of action: Entactogen, euphoriant, psychotomimetic, aphrodisiac
Human active dose: 60-100 mg
Duration of action: 3-5 hours
Toxic manifestations: Nausea, dehydration
Toxicity: Not reported

Although the chemical structure of this drug differs substantially from that of MDMA it has been described as one of the most suitable substitutes for this amphetamine. Interestingly, this compound also produces an effect similar to 4-bromo-2,5-dimethoxyphenethylamine (2CB) on sensorial perception in man [1]. The substance may be easily prepared [2] from a readily available and uncontrolled precursor (hydroquinone). The HPLC properties of this substance have been described [3].

References:

[1] Shulgin, A.T. and Shulgin, A.: PIHKAL, Transform Press, Inc., P.O.Box 13675, Berkeley, CA, p.554, (1991)
[2] Nichols, D.E.; Shulgin, A.T.: J. Pharm. Sci., **65**, p.1554, (1976)
[3] Markovich, R. J.; Qiu, X.; Nichols, D.E.; Pidgeon, C.; Invergo, Ben; Alvarez, Francisco M.: Anal. Chem., **63**, p.1851, (1991)

Substance Number 11 in Chapter 2.1.2

IUPAC Name: 2-(4-Allyloxy-3,5-dimethoxy-phenyl)-ethylamine
Synonyms: 2-(4-Allyloxy-3,5-dimethoxyphenyl)ethanamine
4-Allyloxy-3,5-dimethoxyphenethylamine

Chemical structure:

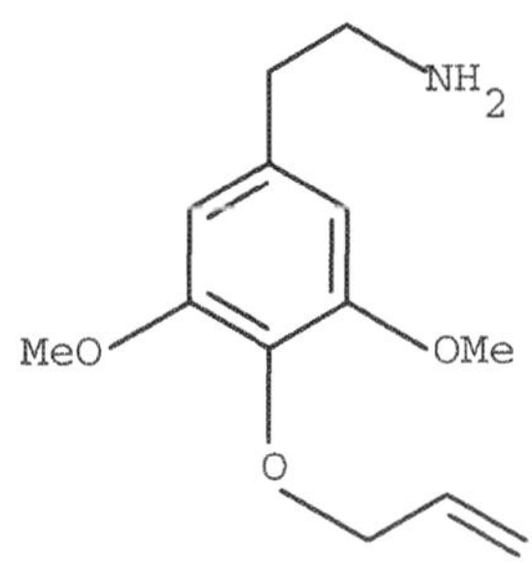

MF: C13H19NO3
MW: 237.321
CA Registry Number: [39201-75-7]
CA Chemical Name: Benzeneethanamine, 2,5-dimethoxy-4-allyloxy-
Category: Psychotomimetic phenethylamine
Street Names: AL
Abuse: Rare
Type of action: Psychotomimetic, entactogen, euphoriant
Human active dose: 20-35 mg
Duration of action: 9-12 hours
Toxic manifestations: Cardiovascular disturbances, nausea, insomnia, confusion, dehydration
Toxicity: Not reported

In man, this phenethylamine has been reported to produce effects similar to those of MDMA but more intense and longer lasting [1,2]. The key-precursor for its manufacture (syringonitrile) has become readily available. The IR spectrum of this substance has been published [1].

References:

[1] Leminger, O.: Chem. Pr., **22**, p.553, (1972)
[2] Shulgin, A.T. and Shulgin, A.: PIHKAL, Transform Press, Inc., P.O.Box 13675, Berkeley, CA, p.460, (1991)

2.1.3 Occasional MDMA Substitutes

The substances described in this chapter are usually sold under the street names typical for MDMA and its derivatives. Being practically devoid of entactogenic activity and often quite toxic, these drugs usually make only poor MDMA substitutes. Consequently, they are recognized and rejected by experienced users.

The two following compounds (**1, 2**) have appeared on the clandestine market because of a nomenclature error [1]. Animal studies show that they possess a trace of MDMA-like activity and are more toxic than MDMA [2,3].

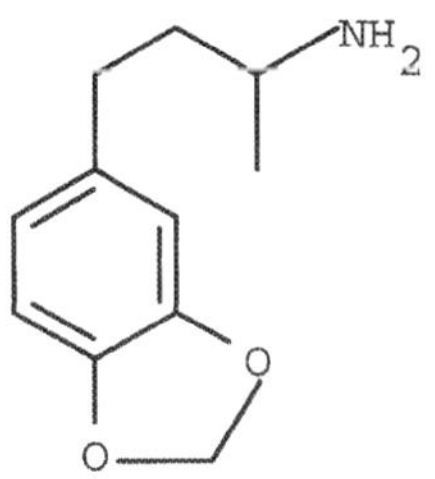

1

200 mg

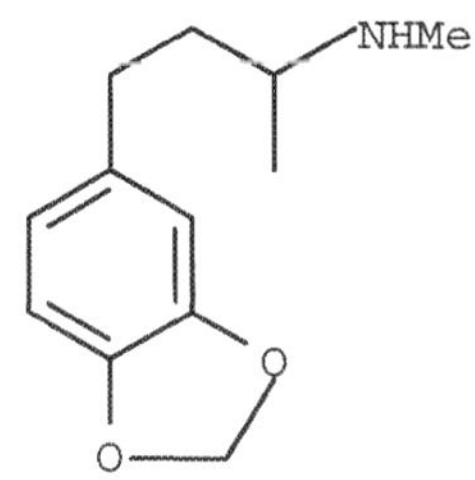

2

200 mg

The substituted 3-phenylpropylamine (**3**) is an active impurity sometimes found in street samples of MDMA. It may be formed, under particular conditions (the presence of peroxides), when MDMA is manufactured from safrole via HBr addition. This frequently used procedure has been thoroughly described on the Internet [4].

The phenethylamine (**4**) has been used for several years in California. Its synthesis has been repeatedly described [5].

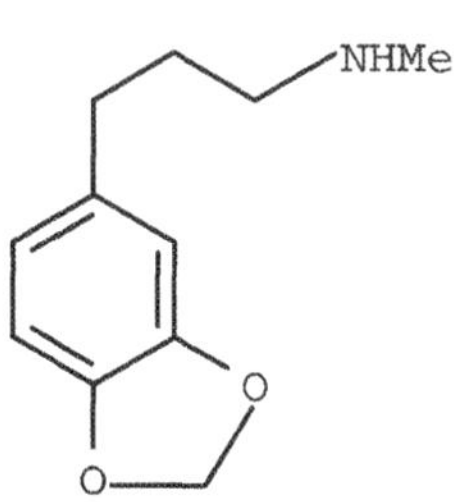

3

200-240 mg

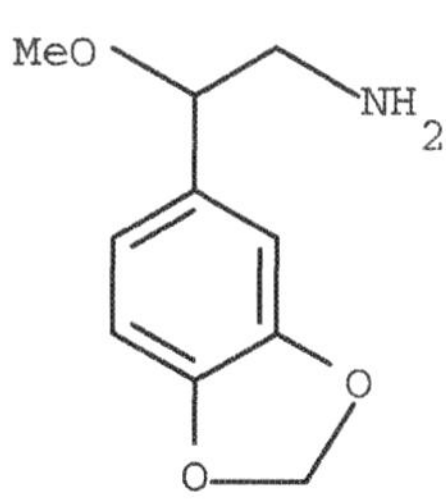

4

80-140 mg

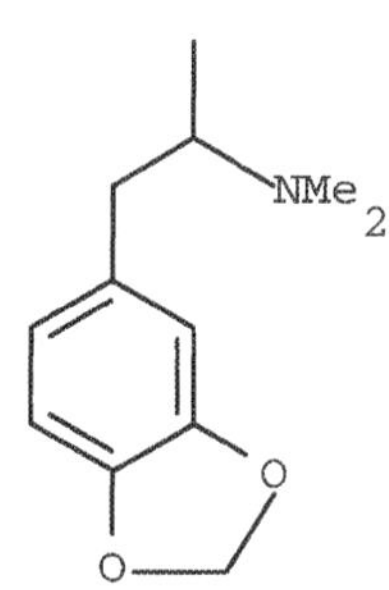

5

400-500 mg

Similarly, the following four amphetamines (**5, 6, 7** and **8**) have been sold as MDMA or MDA. They make only very poor substitutes for these latter drugs. Particularly the last three compounds represent the most dangerous substances in this chapter because their CNS activity appears only at highly toxic doses [6,7].

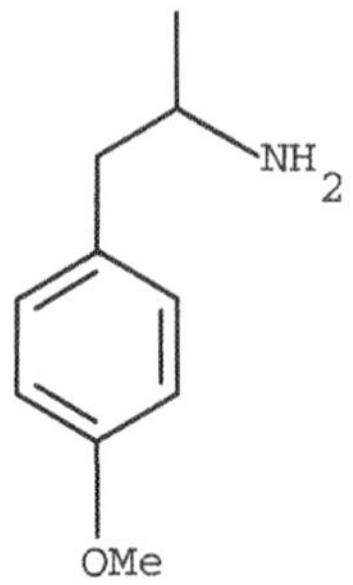

6

50-80 mg

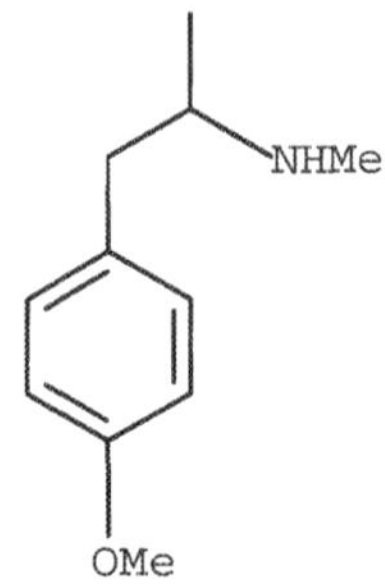

7

100-110 mg

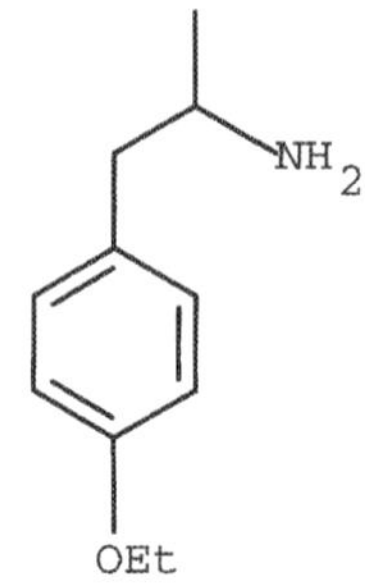

8

Consequently, a number of lethal overdoses caused by these drugs has been reported in the United States and Canada.

2.1.3.1 General References

[1] Shulgin, A.T. et al.: J. Anal. Toxicol., **6,** p.71, (1982)
[2] DeRuiter, J. et al.: Brain Res. Bull., **34**, p.143, (1994)
[3] Davis, W.M.; Borne, R.F.: Substance and Alcohol Actions/Misuse **5**, p.105, (1984)
[4] http://www.damicon.fi/drugs/mdma/ mdma.synth
[5] Torres, M.A.; Cassels, B.; Rezende, M.C.: Synth. Commun., **25,** (8), p.1239, (1995)
[6] Noggle, F.T.; Clark, C.R.; DeRuiter, J.: J. Chromatogr. Sci., **27,** p.602, (1989)
[7] Noggle, F.T.; Clark, C.R.; DeRuiter, J.: J. Chromatogr. Sci., **27,** p.607, (1989)

2.1.3.2 Data Sheets: Occasional MDMA Substitutes

Substance Number 1 in Chapter 2.1.3

IUPAC Name: 3-Benzo[1,3]dioxol-5-yl-1-methyl-propylamine
Synonyms: 1-(3,4-Methylenedioxyphenyl)-3-butanamine
3-Amino-1-(3,4-methylenedioxyphenyl)butane

Chemical structure:

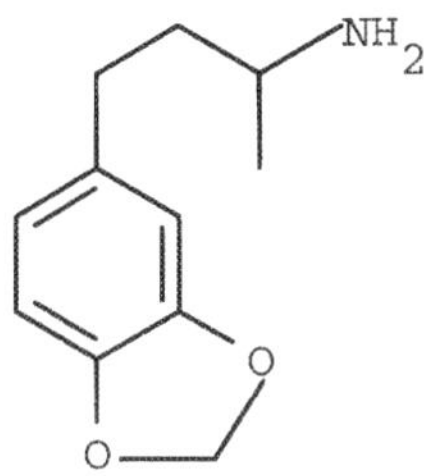

MF: C11H15NO2
MW: 193.245
CA Registry Number: [40742-32-3]; racemate [92279-86-2]
CA Chemical Name: 1,3-Benzodioxole-5-propanamine, α-methyl-
Category: Psychotomimetic phenethylamine
Street Names: The same as for MDA or MDMA; HMDA
Abuse: Rare
Type of action: Stimulant, euphoriant
Human active dose: 200 mg
Duration of action: Short
Toxic manifestations: Cardiovascular disturbances, nausea, hyperthermia, dehydration
Toxicity: Animal studies show this compound is more toxic than MDMA [1].

This drug has appeared on the drug market because of the use of false precursors in MDMA synthesis [3]. In fact, the true MDMA precursor (piperonyl methyl ketone) is sometimes incorrectly marketed as piperonylacetone which is actually a homologous ketone [2]. Consequently, this last compound gives the MDA homologue when subjected to reductive amination [2]. The drug is only a poor MDMA (or MDA) substitute. The identification of this substance by MS and HPLC has been described [3].

References:

[1] Davis, W.M.; Borne, R.F.: Substance and Alcohol Actions/Misuse, **5**, p.105, (1984)
[2] Shulgin, A.T.; Jacob III, P.: J. Anal. Toxicol., **6**, p.71, (1982)
[3] Noggle, F.T.; Clark, C.R.; DeRuiter, J.: J. Chromatogr. Sci., **27**, p.240, (1989)

Substance Number 2 in Chapter 2.1.3

IUPAC Name: (3-Benzo[1,3]dioxol-5-yl-1-methyl-propyl)-methyl-amine
Synonyms: N-Methyl-1-(3,4-methylenedioxyphenyl)-3-butanamine
3-(Methylamino)-1-(3,4-methylenedioxyphenyl)butane

Chemical structure:

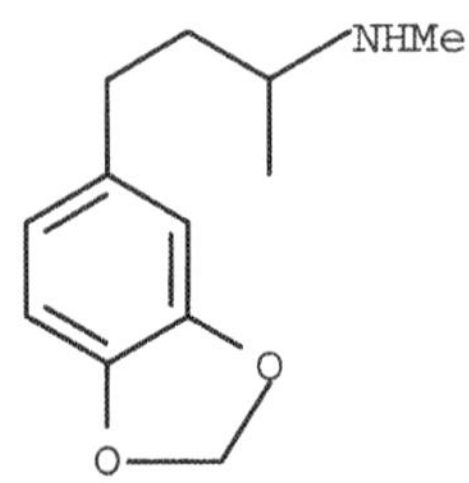

MF: C12H17NO2
MW: 207.272
CA Registry Number: [108248-08-4]; racemate [92279-87-3]
CA Chemical Name: 1,3-Benzodioxole-5-propanamine, N,α-dimethyl-
Category: Psychotomimetic phenethylamine
Street Names: The same as for MDMA; HMDMA
Abuse: Rare
Type of action: Stimulant, euphoriant
Human active dose: 200 mg
Duration of action: Short
Toxic manifestations: Cardiovascular disturbances, nausea, hyperthermia, dehydration
Toxicity: Animal studies show this compound is more toxic than MDMA [1].

This drug has appeared on the drug market because of the use of false precursors in MDMA synthesis [3]. In fact, the true MDMA precursor (piperonyl methyl ketone) is sometimes incorrectly marketed as piperonylacetone which is actually a homologous ketone [2]. Consequently, this last compound gives the MDMA homologue when subjected to reductive amination [2]. The drug is only a poor MDMA substitute. The identification of this substance by MS and HPLC has been described [3].

References:

[1] Davis, W.M.; Borne, R.F.: Substance and Alcohol Actions/Misuse **5**, p.105, (1984)
[2] Shulgin, A.T.; Jacob III, P.: J. Anal. Toxicol., **6,** p.71, (1982)
[3] Noggle, F.T.; Clark, C.R.; DeRuiter, J.: J. Chromatogr. Sci., **27**, p.240, (1989)

Substance Number 3 in Chapter 2.1.3

IUPAC Name: (3-Benzo[1,3]dioxol-5-yl-propyl)-methyl-amine
Synonyms: N-Methyl-(3,4-methylenedioxyphenyl)-3-propanamine
1-(Methylamino)-3-(3,4-methylenedioxyphenyl)propane
iso-MDMA

Chemical structure:

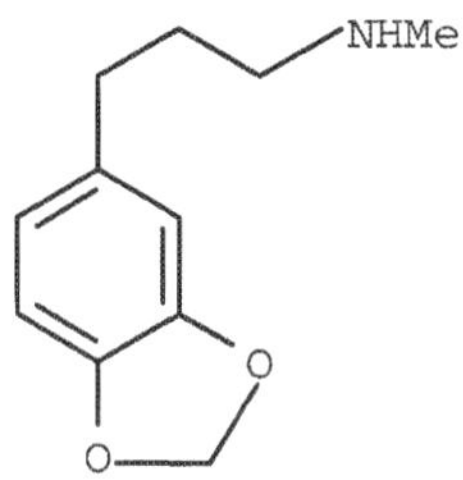

MF: $C_{11}H_{15}NO_2$
MW: 193.245
CA Registry Number: [33543-11-2]
CA Chemical Name: 1,3-Benzodioxole-5-propanamine, N-methyl-
Category: Psychotomimetic phenethylamine
Street Names: GAMMA
Abuse: Rare
Type of action: Euphoriant, psychotomimetic
Human active dose: 200 240 mg
Duration of action: 4 hours
Toxic manifestations: Not reported
Toxicity: Not reported

This MDMA homologue [1] has been found, sometimes in a high proportion, in street samples of MDMA. The compound has been reported to have some MDMA-like activity [2]. The identification of this substance by MS and HPLC has been described [1].

References:

[1] Noggle, F.T.; Clark, C.R.; DeRuiter, J.: Microgram, XXIV, (5) p.114, (1991)
[2] Shulgin, A.T. and Shulgin, A.: PIHKAL, Transform Press, Inc., P.O.Box 13675, Berkeley, CA, p.718, (1991)

Substance Number 4 in Chapter 2.1.3

IUPAC Name: 2-Benzo[1,3]dioxol-5-yl-2-methoxy-ethylamine
Synonyms: 2-Methoxy-2-(3,4-methylenedioxyphenyl)ethanamine
β-Methoxyhomopiperonylamine

Chemical structure:

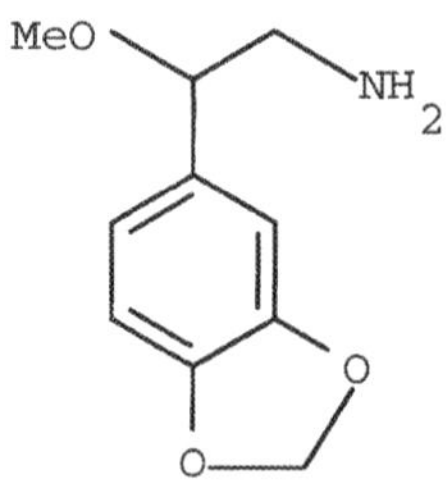

MF: C10H13NO3
MW: 195.218
CA Registry Number: [73304-06-0]; hydrochloride [98537-37-2]
CA Chemical Name: 1,3-Benzodioxole-5-ethanamine, β-methoxy-
Category: Psychotomimetic phenethylamine
Street Names: BOH
Abuse: Rare
Type of action: Euphoriant, psychotomimetic, entactogen
Human active dose: 80-140 mg
Duration of action: 6-8 hours
Toxic manifestations: Nausea, anorexia, confusion, hypertension, cardiovascular disturbances, dehydration
Toxicity: Not reported

This drug is one of the occasional MDMA substitutes. Being more easily manufactured than MDMA itself [1], it is likely to be abused more frequently in the future. In the low dose range (30-40 mg) it is often used as a « mood enhancer » [2].

References:

[1] Torres, M.A.; Cassels, B.; Rezende, M.C.: Synth. Commun., **25**(8), p.1239, (1995)
[2] Shulgin, A.T. and Shulgin, A.: PIHKAL, Transform Press, Inc., P.O.Box 13675, Berkeley, CA, p.496, (1991)

Substance Number 5 in Chapter 2.1.3

IUPAC Name: (2-Benzo[1,3]dioxol-5-yl-1-methyl-ethyl)-dimethyl-amine
Synonyms: N,N-Dimethyl-1-(3,4-methylenedioxyphenyl)-2-propanamine
N-Methyl-MDMA
N,N-Dimethyl-3,4-methylenedioxyamphetamine
1-Dimethylamino-2-(3,4-methylenedioxyphenyl)propane
N,N,α-Trimethyl-3,4-methylenedioxyphenethylamine
N,N-Dimethyl-3,4-methylenedioxyphenylisopropylamine

Chemical structure:

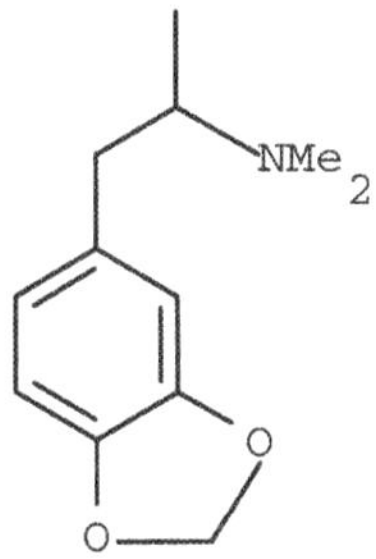

MF: C12H17NO2
MW: 207.272
CA Registry Number: [74698-50-3]; racemate [131206-60-5]
CA Chemical Name: 1,3-Benzodioxol-5-ethanamine, N,N,α-trimethyl-
Category: Psychotomimetic phenethylamine
Street Names: MDDM; MDMMA
Abuse: Rare
Type of action: Euphoriant
Human active dose: 400-500 mg
Duration of action: 4-6 hours
Toxic manifestations: Absence of the libido and nausea has been reported.
Toxicity: In man, a dose of 1000 mg has been tolerated without any important adverse effect [1].

This drug is an occasional and poor MDMA substitute [2,3]. The identification of this drug by GC-MS, NMR [3] and HPLC [4] has been described.

References:

[1] Shulgin, A.T. and Shulgin, A.: PIHKAL, Transform Press, Inc., P.O.Box 13675, Berkeley, CA, p.725, (1991)
[2] Microgram, XXI, (7), p.115, (1988);
[3] Bovolenta, A.; Morselli.O.: Microgram, XXX, (1), p.14, (1997)
[4] Clark, C.R.; Noggle, F.T.; DeRuiter, J.: J. Liq. Chromatogr., **13**, p.263, (1990)

Substance Number 6 in Chapter 2.1.3

IUPAC Name: 2-(4-Methoxy-phenyl)-1-methyl-ethylamine
Synonyms: 1-(4-Methoxyphenyl)-2-propanamine
4-Methoxyamphetamine
α-Methyl-4-methoxyphenethylamine
4-Methoxyphenylisopropylamine
1-Amino-2-(4-methoxyphenyl)propane

Chemical structure:

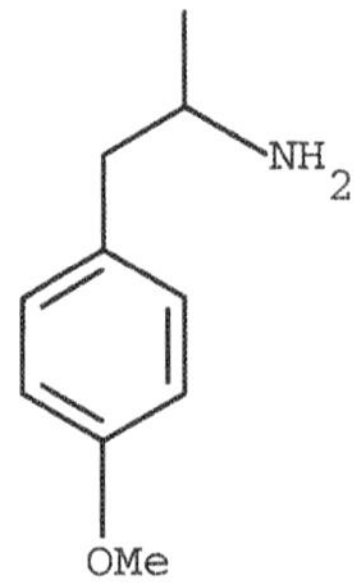

MF: C10H15NO
MW: 165.235
CA Registry Number: [64-13-1]; racemate [23239-32-9]; [R]-58993-79-6]; [S]-[58993-78-5]; hydrochloride [R]-[50505-80-1]
CA Chemical Name: Benzeneethanamine, 4-methoxy-α-methyl-
Category: Psychotomimetic phenethylamine
Street Names: PMA; Chicken Power
Abuse: Limited
Type of action: Stimulant, euphoriant
Human active dose: 50-80 mg
Duration of action: 2-3 hours
Toxic manifestations: Pronounced hypertension, cardiovascular disturbances, nausea, anxiety, confusion, dehydration
Toxicity: Several hundreds milligrams of this substance may be a lethal dose in man.

This drug is considered to be one of the most dangerous amphetamines. Being sold as MDA or MDMA, its absorption may lead to lethal intoxication.

Many such cases have been reported in the United States and Canada [2]. The identification of this drug by MS and HPLC has been described [2].

References:

[1] Shulgin, A.T. and Shulgin, A.: PIHKAL, Transform Press, Inc., P.O.Box 13675, Berkeley, CA, p.708, (1991)
[2] Noggle, F.T.; Clark, C.R.; DeRuiter, J.: J. Chromatogr. Sci., **27,** p.602, (1989)

Substance Number 7 in Chapter 2.1.3

IUPAC Name: 2-(4-Methoxy-phenyl-1-methyl-ethyl)-methyl-amine
Synonyms: 1-(4-Methoxyphenyl)-N-methyl-2-propanamine
4-Methoxymethamphetamine
N,α-Dimethyl-4-methoxyphenethylamine
4-Methoxyphenyl-N-methylisopropylamine
1-(4-Methoxyphenyl)-2-methylaminopropane

Chemical structure:

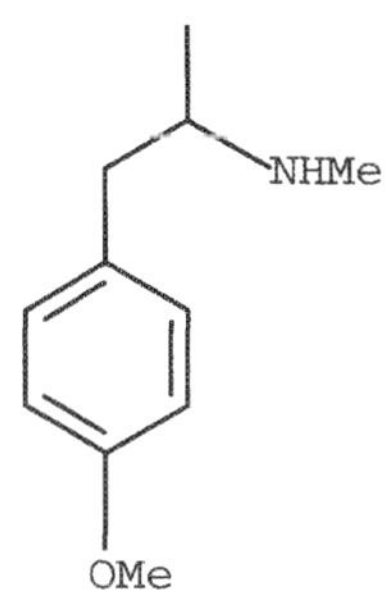

MF: C11H17NO
MW: 179.261
CA Registry Number: [22331-70-0]; racemate [88543-18-4]; hydrochloride-[118931-17-2]
[S]-[113429-54-2]
CA Chemical Name: Benzeneethanamine, 4-methoxy-N,α-dimethyl-
Category: Psychotomimetic phenethylamine
Street Names: PMMA; 4-MMA
Abuse: Limited
Type of action: Stimulant, euphoriant
Human active dose: 100-110 mg
Duration of action: 2-3 hours
Toxic manifestations: Pronounced hypertension, cardiovascular disturbances
Toxicity: Several hundreds milligrams of this substance may be a lethal dose in man [1].

This drug is considered to be one of the most dangerous amphetamines. Being usually sold as MDA or MDMA, its absorption may lead to lethal intoxication. Many such cases have been reported in the United States and Canada [1]. This reference also describes the synthesis and identification of this drug by MS and HPLC. The substance was reported to have MDMA-like properties in animal studies [2], but studies in man [3] have not confirmed these findings.

References:

[1] Noggle, F.T.; Clark, C.R.; DeRuiter, J.: J. Chromatogr. Sci., **27,** p.602, (1989)
[2] Glennon, R.A.; Higgs, R.: Pharmacol. Biochem. Behav., **46**, p.759, (1992)
[3] Shulgin, A.T. and Shulgin, A.: PIHKAL, Transform Press, Inc., P.O.Box 13675, Berkeley, CA, p.784, (1991)

Substance Number 8 in Chapter 2.1.3

IUPAC Name: 2-(4-Ethoxy-phenyl)-1-methyl-ethylamine
Synonyms: 1-(4-Ethoxyphenyl)-2-propanamine
4-Ethoxyamphetamine
α-Methyl-4-methoxyphenethylamine
4-Ethoxyphenylisopropylamine
1-Amino-2-(4-ethoxyphenyl)propane

Chemical structure:

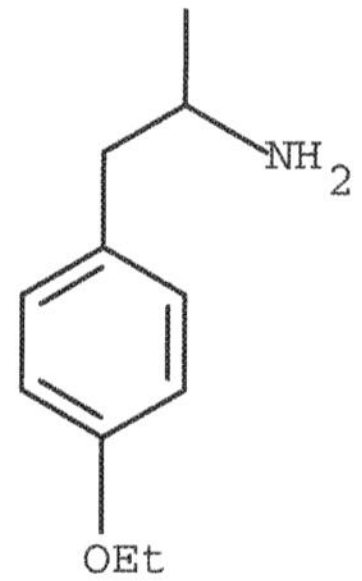

MF: C11H17NO
MW: 179.261
CA Registry Number: [135014-85-6]; racemate [129476-58-0]
CA Chemical Name: Benzeneethanamine, 4-ethoxy-α-methyl-
Category: Psychotomimetic phenethylamine
Street Names: 4-EA
Abuse: Limited
Type of action: Stimulant, euphoriant
Human active dose: Not reported
Duration of action: Not reported
Toxic manifestations: Marked hypertension, cardiovascular disturbances
Toxicity: Several hundreds milligrams of this substance may be a lethal dose in man.

This drug is considered to be one of the most dangerous amphetamines [1]. Usually sold as MDA or MDMA, its absorption may lead to lethal intoxication. Many such cases have been reported in the United States and Canada [1,2]. The first reference also describes the synthesis and identification of this drug by MS and HPLC.

References:

[1] By, A.W.; Duhaime, R.; Lodge, B.A.: Forensic Sci. Int., **49**, p.162, (1991)
[2] Microgram, XXI, (8), p.115, (1988)

2.1.4 Other Psychotomimetic Phenethylamines

The compounds included in this chapter are usually referred to as hallucinogens because of the pronounced visual effects they produce in man. In this sense, their activity resembles that of LSD, but in the case of phenethylamines, prominent euphoria is more frequently reported as a component of their action. The most frequent physiological effects include tachycardia, hypertension, mydriasis, hyperthermia, tremor, hyperreflexia and nausea.

Two compounds in this class (DOM, **1**, « STP » and DOB, **5**) [1] were widely abused in the 1970s. Abuse of their analogues (**2**, **3**, **4**, **5**) [2], which also show clear hallucinogenic properties, is less frequent.

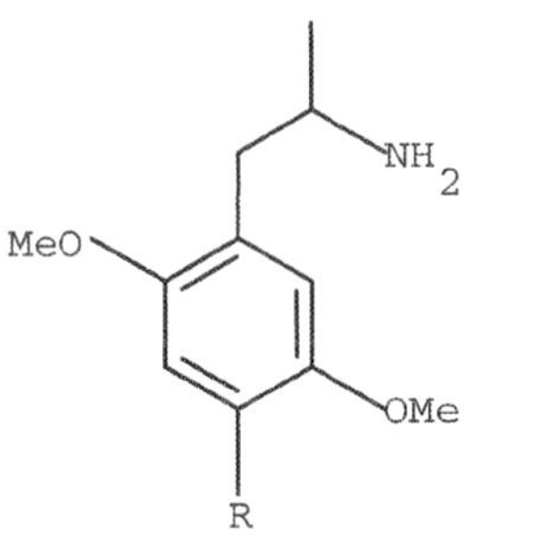

1, DOM, R=Me 3-10 mg
2, DOET, R=Et 2-6 mg
3, DOPR, R=Pr 3-5 mg

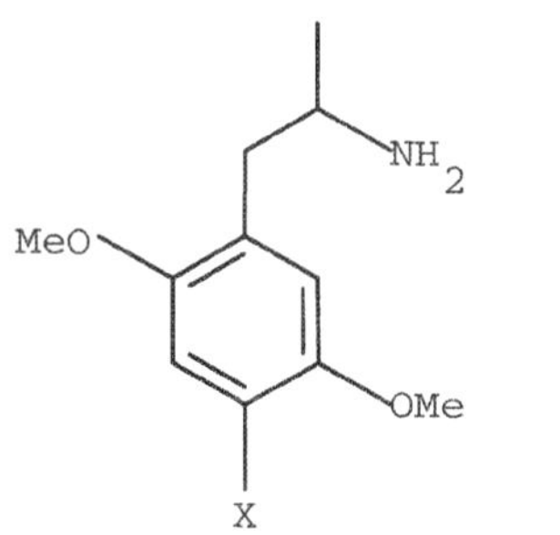

4, DOC, X=Cl 1.5-3 mg
5, DOB, X=Br 1-3 mg
6, DOI, X-I 1.5-3 mg

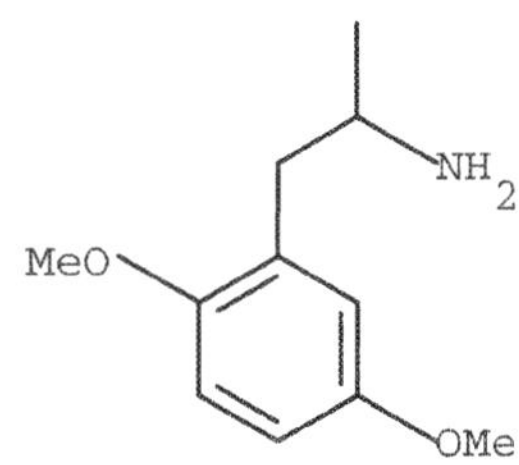

7
80-160 mg

The three amphetamines (**4**, **5**, **6**) may be easily prepared by simple halogenation from 2,5-dimethoxyamphetamine (**7**) which is a widely used chemical in the pharmaceutical and photographic [3] industries.

Similarly to LSD, these drugs are usually sold absorbed on a piece of blotter paper, sugar cubes etc.

These drugs show a particularly long latency of action (2-3 hours) which may be potentially dangerous to an inexperienced user. Thus, hundreds of serious overdoses occurred in the USA in 1967 [4] after the free and abundant distribution of capsules containing 20 mg of DOM.

Some impatient users who had experience with rapidly acting LSD took 2 or even 3 such capsules and were consequently seriously intoxicated for several days.

There are other potent hallucinogenic phenethylamines such as TMA2 (**8**) [5], its ethyl analogue (MEM, **9**) and the methylenedioxy analogues (**10**, **11**).

Seizures of the last three compounds in Canada have recently been reported [6].

The following powerful compounds (Aleph1, **12**, and Aleph2, **13**) may be considered as thioanalogues of TMA2 and MEM, respectively [7].

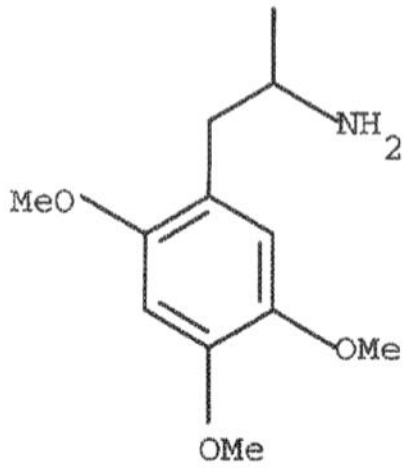
8
20-40 mg

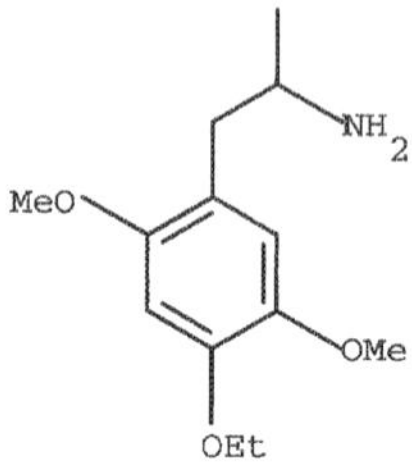
9
20-50 mg

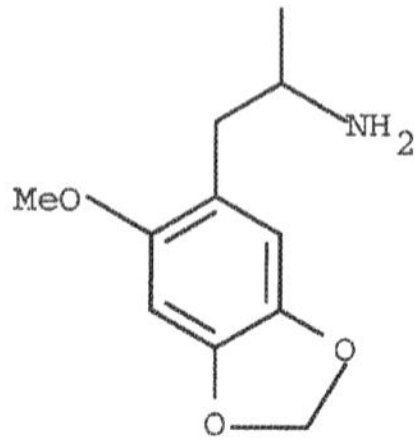
10
25-50 mg

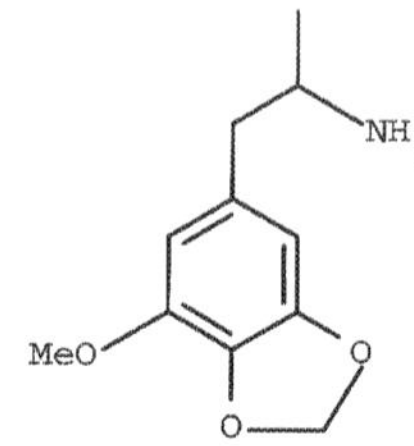
11
100-250 mg

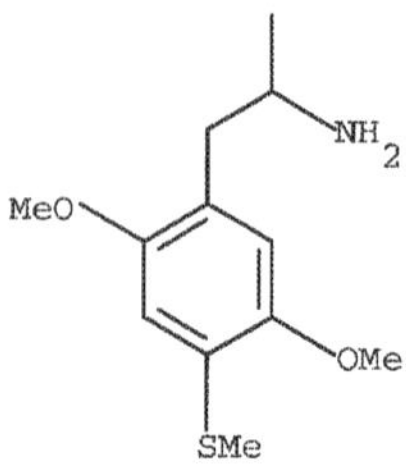
12
5-10 mg

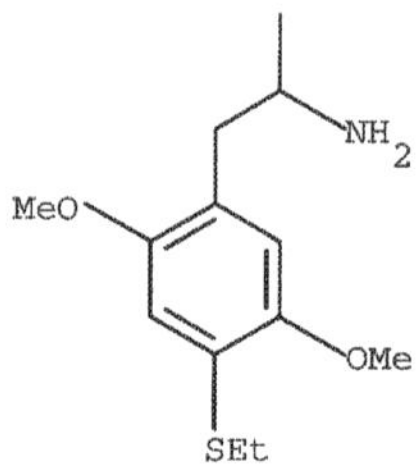
13
4-8 mg

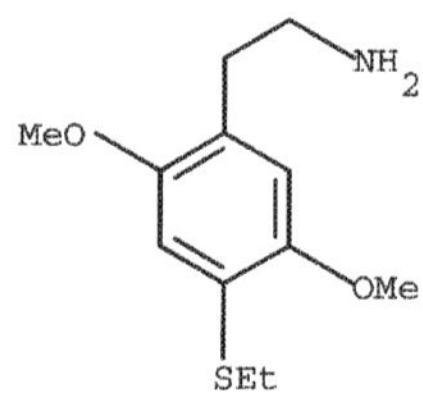
14
12-25 mg

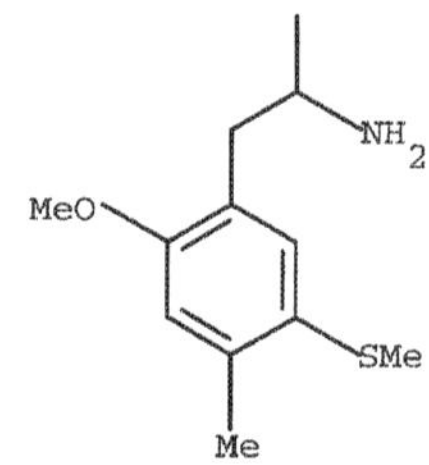
15
30-50 mg

In man, these substances have been reported to produce pronounced euphoria with variable visual hallucinatory activity, although the phenethylamine analogue of (**13**), the substance (**14**) is a classical hallucinogen of high abuse potential. These drugs can be prepared from hydroquinone, which is also the key-precursor for most of the compounds described in this chapter.

This substance (**15**) which is the 5-thioanalogue of DOM has recently been seized in Canada.

The last three phenethylamines in this chapter (**16**, **17**, **18**) form a small subclass among psychotomimetics because they exhibit a particular effect on sensorial perceptions in man. Increased sexual desire, pleasure and of all the senses such as touch and taste have been reported. In others respects they have the profile of the classical hallucinogenic phenethylamines [8].

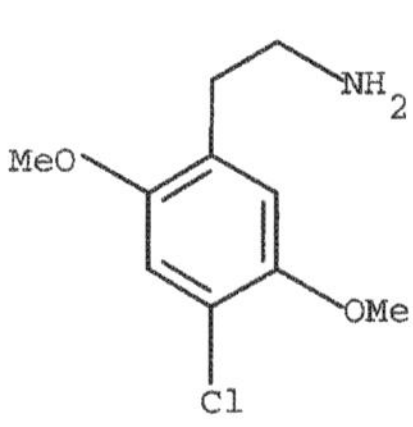
16
20-40 mg

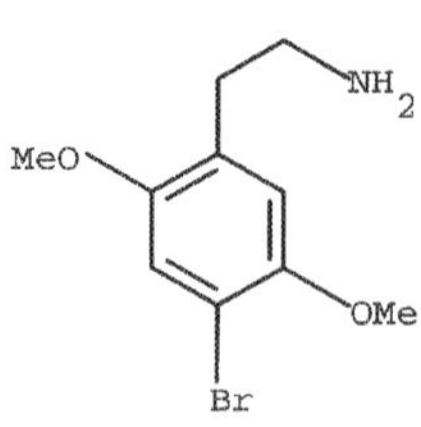
17
12-24 mg

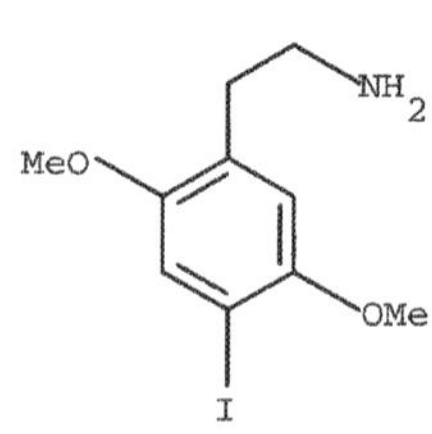
18
14-22 mg

The most popular drug of this subclass is 2CB, (**17**) (« Bromo », « Venus », Nexus », « Spectrum » « BDMPEA »). The abuse of this substance was first reported in 1979 in Texas [8].

Today this drug is frequently found in the « Rave » environment [9]. Also, administration of this substance after an intake of MDMA seems to be very popular [10].

Two analogues (**16** and **18**) share the effects of this drug on sensorial perception, but appear much less frequently in the clandestine market.

2.1.4.1 General References

[1] Stafford, P.: Psychedelic Encyclopedia, Ronin Publishing, Inc., Box 1035, Berkeley, CA, pp.41 and 292, (1992)
[2] McCann, Una; Ricaurte, G.A.: Use and Abuse of Ring Substituted Amphetamines. In: Amphetamine and its analogs; Cho, A.K. and Segal, D.S. (editors); Academic Press, p.371, (Chapter 12), (1994)
[3] Shulgin, A.T.: Psychotomimetic drugs: Structure-Activity Relationships. In: Handbook of Psychopharmacology, Plenum Press, New York, Vol.**11**, p.281, (1978)
[4] Ref. 1, p.291
[5] Ref. 3, p.284
[6] By, A.W. et al.: J. Forensic Sci., **35** (2), p.316, (1990)
[7] Shulgin, A.T. and Shulgin, A.: PIHKAL, Transform Press, Inc., P.O.Box 13675, Berkeley, CA, pp.461 and 462, respectively, (1991)
[8] DeRuiter, J.; Clark, C.R.; Noggle, F.T.: J. Chromatogr. Sci., **33**, p.583, (1995)
[9] http://www.damicon.fi/drugs/psychedelics/phenethylamines/2cb.info
[10] http:// www.ecstasy.org/combinations.html

2.1.4.2 Data Sheets: Other Psychotomimetic Phenethylamines

Substance Number 1 in Chapter 2.1.4

IUPAC Name: 2-(2,5-Dimethoxy-4-methyl-phenyl)-1-methyl-ethylamine
Synonyms: 1-(2,5-Dimethoxy-4-methylphenyl)-2-propanamine
2,5-Dimethoxy-4-methylamphetamine
2,5-Dimethoxy-4,α-dimethylphenethylamine
2,5-Dimethoxy-4-methylphenylisopropylamine
1-Amino-2-(2,5-dimethoxy-4-methylphenyl)propane

Chemical structure:

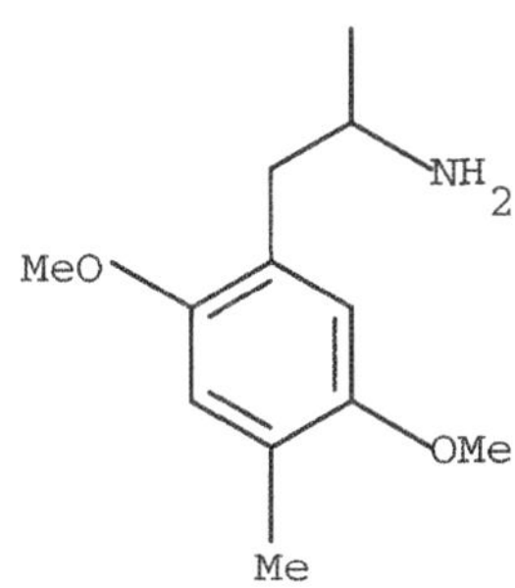

MF: C12H19NO2
MW: 209.288
CA Registry Number: [15588-95-1]; racemate [26011-50-7]; [R]-43061-13-8]; [S]-[43061-14-9]; hydrochloride [15589-00-1]
CA Chemical Name: Benzeneethanamine, 2,5-dimethoxy-α,4-dimethyl-
Category: Psychotomimetic phenethylamine
Street Names: DOM; STP
Abuse: Frequent
Type of action: Psychotomimetic, hallucinogen, euphoriant, stimulant
Human active dose: 3-10 mg
Duration of action: 14-20 hours
Toxic manifestations: Insomnia, anxiety, confusion, cardiovascular disturbances, tachycardia, hypertension, hyperthermia, nausea, tremor
Toxicity: LD50 is 89 mg/kg i.p. (mouse) [1].

The abuse of this amphetamine has become less frequent than in the 1960s [2]. The drug has a very long latency of action (2-3 hours). The identification of this substance by MS and HPLC has been described [3].

References:

[1] Ho, T. Beng, et al.: J. Med. Chem., **13**, p.26, (1970)
[2] McCann, Una; Ricaurte, G.A.: Use and Abuse of Ring-Substituted Amphetamines. In: Amphetamine and its analogs; Cho, A.K. and Segal, D.S. (editors); Academic Press, p.377, (Chapter 12), (1994)
[3] Mills III, T.; Roberson, J.C.; McCurdy, H.H.; Wall, W.H.: Instrumental Data for Drug Analysis, Elsevier Science Publishing Co., Inc., p.730, (1992)

Substance Number 2 in Chapter 2.1.4

IUPAC Name: 2-(4-Ethyl-2,5-dimethoxy-phenyl)-1-methyl-ethylamine
Synonyms: 4-Ethyl-2,5-dimethoxyamphetamine
4-Ethyl-2,5-dimethoxy-α-methylphenethylamine
4-Ethyl-2,5-dimethoxyphenylisopropylamine
1-Amino-2-(4-ethyl-2,5-dimethoxy phenyl)propane

Chemical structure:

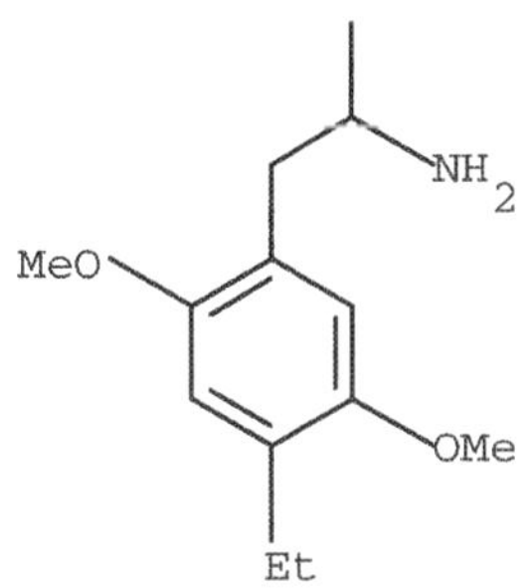

MF: C13H21NO2
MW: 223.314
CA Registry Number: [22004-32-6]; racemate [41538-40-3]; [R]-[57116-37-7]; [S]-[53305-83-2]; hydrochloride, racemate [53581-54-7]
CA Chemical Name: Benzeneethanamine, 4-ethyl-2,5-dimethoxy-α-methyl-
Category: Psychotomimetic phenethylamine
Street Names: DOET
Abuse: Limited
Type of action: Euphoriant, psychotomimetic, hallucinogen, stimulant
Human active dose: 2-6 mg
Duration of action: 14-20 hours
Toxic manifestations: Insomnia, anxiety, confusion, cardiovascular disturbances, dehydration
Toxicity: LD50 is 76 mg/kg i.p.(mouse) [1].

This amphetamine is mostly used in a low dose range (about 1 mg) [2] as an « aesthetic enhancer ». Similarly to its 4-methyl analogue (DOM), it also shows long latency of action. The identification of this drug by MS and HPLC has been described [3].

References:

[1] Ho, T. Beng, et al.: J. Med. Chem., **13**, p.26, (1970)
[2] Shulgin, A.T. and Shulgin, A.: PIHKAL, Transform Press, Inc., P.O.Box 13675, Berkeley, CA, p.633, (1991)
[3] Mills III, T.; Roberson, J.C.; McCurdy, H.H.; Wall, W.H.: Instrumental Data for Drug Analysis, Elsevier Science Publishing Co., Inc., p.728, (1992)

Substance Number 3 in Chapter 2.1.4

IUPAC Name: 2-(2,5-Dimethoxy-4-propyl-phenyl)-1-methyl-ethylamine
Synonyms: 1-(2,5-Dimethoxy-4-propylphenyl)-2-propanamine
2,5-Dimethoxy-4-propylamphetamine
2,5-Dimethoxy-α-methyl-4-propylphenethylamine
2,5-Dimethoxy-4-propylphenylisopropylamine
1-Amino-2-(2,5-dimethoxy-4-propylphenyl)propane

Chemical structure:

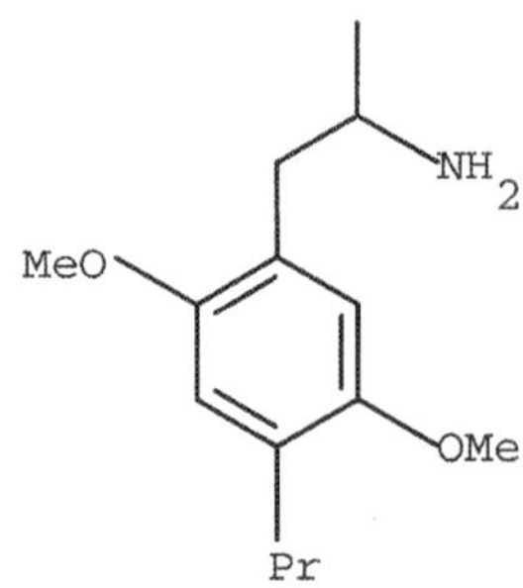

MF: C14H23NO2
MW: 237.341
CA Registry Number: [63779-88-4]
CA Chemical Name: Benzeneethanamine, 2,5-dimethoxy-α-methyl-4-propyl-
Category: Psychotomimetic phenethylamine
Street Names: DOPR
Abuse: Rare
Type of action: Psychotomimetic, hallucinogen, euphoriant, stimulant
Human active dose: 2-5 mg
Duration of action: 20-30 hours
Toxic manifestations: Insomnia, anxiety, confusion, cardiovascular disturbances, dehydration
Toxicity: LD50 is 70 mg/kg i.p. (mouse) [1].

Similarly to its 4-methyl analogue (DOM), this substance also shows long latency of action [2]. The C13-MNR spectrum of this compound has been reported [3].

References:

[1] Ho, T. Beng, et al.: J. Med. Chem., **13**, p.26, (1970)
[2] Shulgin, A.T. and Shulgin, A.: PIHKAL, Transform Press, Inc., P.O.Box 13675, Berkeley, CA, p.648, (1991)
[3] Bienfait, B.: J. Chem. Inf. Comput. Sci., **34**, p.890, (1994)

Substance Number 4 in Chapter 2.1.4

IUPAC Name: 2-(4-Chloro-2,5-dimethoxy-phenyl)-1-methyl-ethylamine
Synonyms: 1-(4-Chloro-2,5-dimethoxyphenyl)-2-propanamine
4-Chloro-2,5-dimethoxyamphetamine
4-Chloro-2,5-dimethoxy-α-methylphenethylamine
4-Chloro-2,5-dimethoxyphenylisopropylamine
1-Amino-2-(4-chloro-2,5-dimethoxyphenyl)propane

Chemical structure:

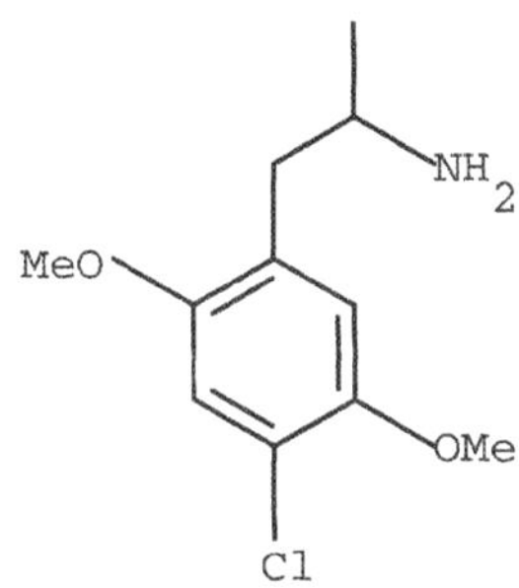

MF: C11H16ClNO2
MW: 229.706
CA Registry Number: [123431-31-2]
CA Chemical Name: Benzeneethanamine, 4-chloro-2,5-dimethoxy-α-methyl-
Category: Psychotomimetic phenethylamine
Street Names: DOC
Abuse: Rare
Type of action: Psychotomimetic, hallucinogen, euphoriant, stimulant
Human active dose: 1.5-3 mg
Duration of action: 12-24 hours
Toxic manifestations: Insomnia, anxiety, confusion, cardiovascular disturbances, dehydration
Toxicity: Not reported

This amphetamine appeared recently on the clandestine market in the USA [1]. It possesses properties similar to its 4-bromo analogue and can also be prepared by chlorination from 2,5-dimethoxyamphetamine. Similar to its 4-methyl analogue (DOM), this substance also shows long latency of action [2].

References:

[1] Microgram, XXI, (7), p.114, (1988)
[2] Shulgin, A.T. and Shulgin, A.: PIHKAL, Transform Press, Inc., P.O.Box 13675, Berkeley, CA, p.622, (1991)

Substance Number 5 in Chapter 2.1.4

IUPAC Name: 2-(4-Bromo-2,5-dimethoxy-phenyl)-1-methyl-ethylamine
Synonyms: 1-(4-Bromo-2,5-dimethoxyphenyl)-2-propanamine
4-Bromo-2,5-dimethoxyamphetamine
4-Bromo-2,5-dimethoxy-α-methylphenethylamine
4-Bromo -2,5-dimethoxyphenylisopropylamine
1-Amino-2-(4-bromo-2,5-dimethoxyphenyl)propane

Chemical structure:

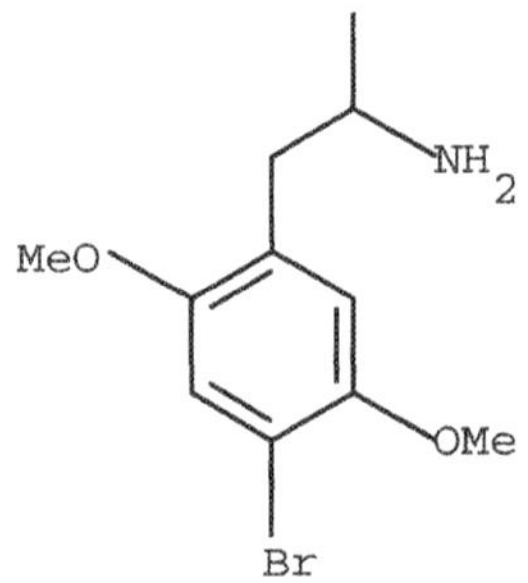

MF: C11H16BrNO2
MW: 274.157
CA Registry Number: [32156-26-6]; racemate [64638-07-9]; [R]-[43061-15-0]; [S]-[43061-16-1]; hydrochloride, racemate [51261-87-1]; hydrochloride [R]-[50505-92-5]; hydrochloride [S]-[50505-93-6]
CA Chemical Name: Benzeneethanamine, 4-bromo-2,5-dimethoxy-α-methyl-
Category: Psychotomimetic phenethylamine
Street Names: DOB
Abuse: Limited
Type of action: Psychotomimetic, hallucinogen, euphoriant, stimulant
Human active dose: 1-3 mg
Duration of action: 18-30 hours
Toxic manifestations: Insomnia, anxiety, confusion, cardiovascular disturbances, dehydration
Toxicity: In man, the LD 50 is estimated to about 100 mg [1].

Similarly to its 4-methyl analogue (DOM) this substance also shows long latency of action [1]. This drug is one of the most potent psychotomimetic phenethylamines known. It may be easily prepared by a simple bromination from 2,5-dimethoxyamphetamine (2,5DMA) which is widely used in the pharmaceutical and photographic industries. Recently, this drug was seized in Germany. The identification of this drug by MS and HPLC has been described [2].

References:

[1] Shulgin, A.T. and Shulgin, A.: PIHKAL, Transform Press, Inc., P.O.Box 13675, Berkeley, CA, p.622, (1991)
[2] Mills III, T.; Roberson, J.C.; McCurdy, H.H.; Wall, W.H.: Instrumental Data for Drug Analysis, Elsevier Science Publishing Co., Inc., p.246, (1992)

Substance Number 6 in Chapter 2.1.4

IUPAC Name: 2-(4-Iodo-2,5-dimethoxy-phenyl)-1-methyl-ethylamine
Synonyms: 1-(4-Iodo-2,5-dimethoxy)-2-propanamine
4-Iodo-2,5-dimethoxyamphetamine
4-Iodo-2,5-dimethoxy -α-methylphenethylamine
4-Iodo-2,5-dimethoxy phenylisopropylamine
1-Amino-2-(4-iodo-2,5-dimethoxy)propane

Chemical structure:

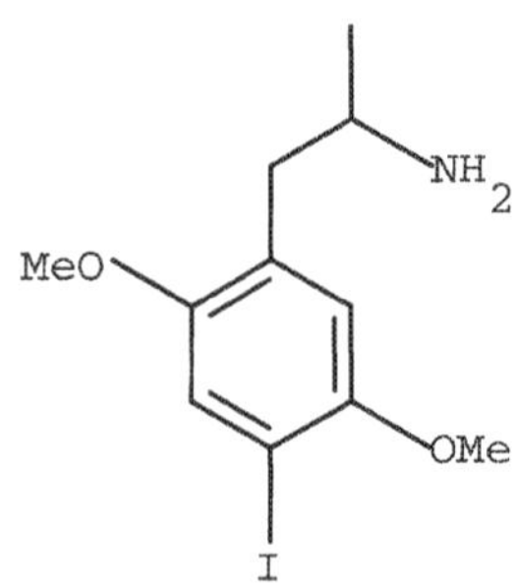

MF: C11H16INO2
MW: 321.157
CA Registry Number: [64584-34-5]; racemate [82830-53-3]; [R]-[82864-06-0]; [S]-[99665-04-0]; hydrochloride [42203-78-1]; hydrochloride, racemate [82830-44-2]; hydrochloride [R]-[82864-02-6]; hydrochloride [S]-[99665-05-1]
CA Chemical Name: Benzeneethanamine, 4-iodo-2,5-dimethoxy-α-methyl-
Category: Psychotomimetic phenethylamine
Street Names: DOI
Abuse: Rare
Type of action: Psychotomimetic, hallucinogen, euphoriant, stimulant
Human active dose: 1.5-3 mg
Duration of action: 16-30 hours
Toxic manifestations: Insomnia, anxiety, confusion, cardiovascular disturbances, dehydration
Toxicity: Not reported

It possesses properties similar to its 4-bromo analogue and it can also be prepared by iodination from 2,5-dimethoxyamphetamine. Similarly to its 4-methyl analogue (DOM), this substance also shows long latency of action [1]. This substance marked by radioactive iodine is frequently employed in brain neurotransmitter research. The HPLC properties of this substance have been studied [2].

References:

[1] Shulgin, A.T. and Shulgin, A.: PIHKAL, Transform Press, Inc., P.O.Box 13675, Berkeley, CA, p.633, (1991)
[2] Biagi, G. L. et al.: J. Chromatogr., A, **723**, p.135, (1996)

Substance Number 7 in Chapter 2.1.4

IUPAC Name: 2-(2,5-Dimethoxy-phenyl)-1-methyl-ethylamine
Synonyms: 1-(2,5-Dimethoxyphenyl)-2-propanamine
2,5-Dimethoxyamphetamine
2,5-Dimethoxy-α-methylphenethylamine
2,5-Dimethoxyphenylisopropylamine
1-Amino-2-(2,5-dimethoxyphenyl)propane

Chemical structure:

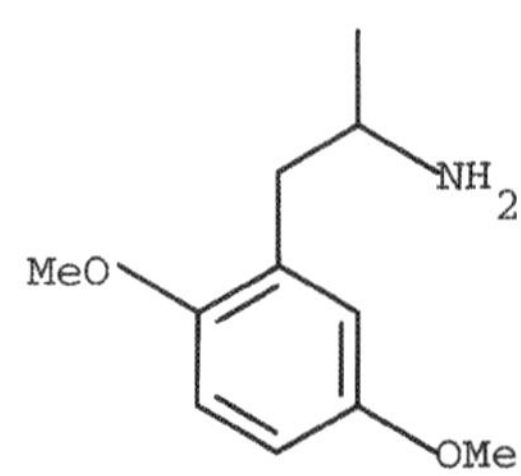

MF: C11H17NO2
MW: 195.261
CA Registry Number: [2801-68-5]; racemate [13641-74-2]; [R]-[58993-81-0]; [S]-[58993-80-9]; hydrochloride [24973-25-9]
CA Chemical Name: Benzeneethanamine, 2,5-dimethoxy-α-methyl-
Category: Psychotomimetic phenethylamine
Street Names: 2,5DMA
Abuse: Limited
Type of action: Stimulant, euphoriant, hallucinogen
Human active dose: 80-160 mg
Duration of action: 6-8 hours
Toxic manifestations: Pronounced hypertension, cardiovascular disturbances, nausea, anxiety, confusion, dehydration
Toxicity: LD50 is 135 mg/kg i.p. (mouse) [1].

This substance is an occasional and relatively poor MDMA substitute [2, 3]. It is a particularly important amphetamine because it can be converted into very potent psychotomimetics (DOB, DOC and DOI) by simple halogenation. The identification of this drug by MS and HPLC has been described [4].

References:

[1] Ho, T. Beng, et al.: J. Med. Chem., **13**, p.26, (1970)
[2] Shulgin, A.T.: Psychotomimetic drugs: Structure-Activity Relationships. In: Handbook of Psychopharmacology, Plenum Press, New York, Vol.**11**, p.281, (1978)
[3] Cassels, B.K.; Gomez-Jeria, S.J.: J. of Psychoactive Drugs, **17**, p.129, (1985)
[4] Mills III, T.; Roberson, J.C.; McCurdy, H.H.; Wall, W.H.: Instrumental Data for Drug Analysis, Elsevier Science Publishing Co., Inc., p.720, (1992)

Substance Number 8 in Chapter 2.1.4

IUPAC Name: 1-Methyl-2-(2,4,5-trimethoxy-phenyl)-ethylamine
Synonyms: 1-(2,4,5-Trimethoxyphenyl)-2-propanamine
2,4,5-Trimethoxyamphetamine
α-Methyl-2,4,5-trimethoxyphenethylamine
2,4,5-Trimethoxyphenylisopropylamine
1-Amino-2-(2,4,5-trimethoxyphenyl)propane

Chemical structure:

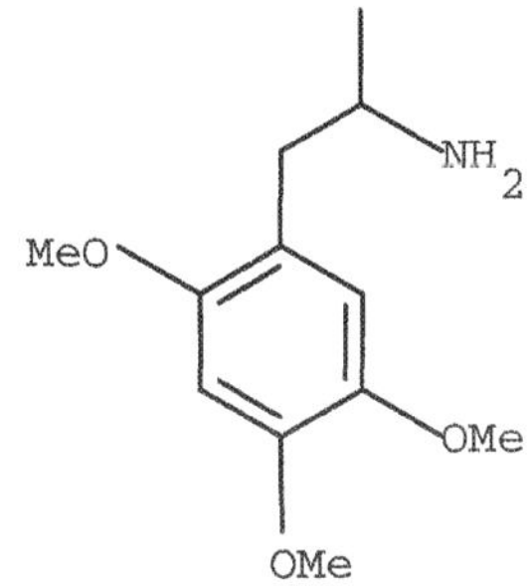

MF: $C_{12}H_{19}NO_3$
MW: 225.287
CA Registry Number: [1083-09-6]; racemate [22199-15-1]
CA Chemical Name: Benzeneethanamine, 2,4,5-trimethoxy-α-methyl-
Category: Psychotomimetic phenethylamine
Street Names: TMA2
Abuse: Rare
Type of action: Psychotomimetic, hallucinogen, euphoriant, stimulant
Human active dose: 20-40 mg
Duration of action: 8-12 hours
Toxic manifestations: Insomnia, anxiety, confusion, cardiovascular disturbances, dehydration
Toxicity: LD50 is 180 mg/kg i.p. (mouse) [1].

Recently, this amphetamine was seized on the « Rave » scene in Germany. Its abuse may become more frequent because it can be prepared from a readily available precursor (benzoquinone). Also, a simple conversion of asarone obtained from natural sources (oil of Calamus root from Acorus calamus L.) into this amphetamine has been described in the clandestine literature [2]. The NMR spectrum [3] of this substance and its identification by GC-ECD [4] and HPLC [5] have been published.

References:

[1] Ho, T. Beng, et al.: J. Med. Chem., **13**, p.26, (1970)
[2] Uncle Fester: Practical LSD Manufacture, Loompanics Unlimited, Port Townsend, WA 98368, p.77, (1995)
[3] Bailey, K. et al.: Can. J. Chem., **49**, p.3143, (1971)
[4] Midha, K. K.; Cooper, J. K.; Gagne, D.; Bailey, K.: J. Anal. Toxicol., **3**, p.53, (1979)
[5] Markovich, R. J.; Qiu, X.; Nichols, D. E.; Pidgeon, C.; Invergo, Ben; Alvarez, Francisco M.: Anal. Chem., **63**, p.1851, (1991)

Substance Number 9 in Chapter 2.1.4

IUPAC Name: 2-(4-Ethoxy-2,5-dimethoxy-phenyl)-1-methyl-ethylamine
Synonyms: 1-(4-Ethoxy-2,5-dimethoxyphenyl)-2-propanamine
4-Ethoxy-2,5-dimethoxyamphetamine
4-Ethoxy-2,5-dimethoxy-α-methylphenethylamine
4-Ethoxy-2,5-dimethoxyphenylisopropylamine
1-Amino-2-(4-ethoxy-2,5-dimethoxyphenyl)propane

Chemical structure:

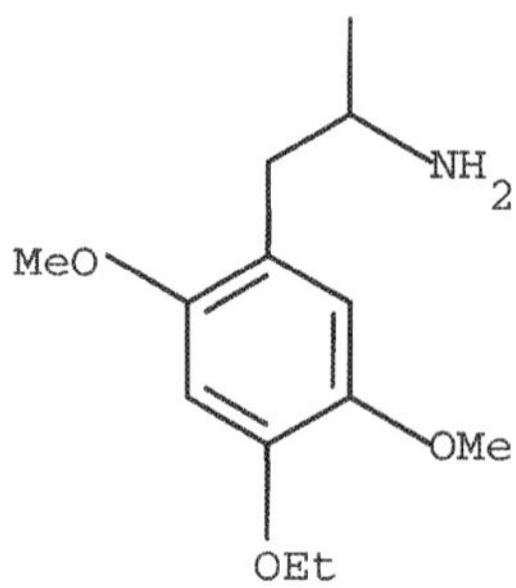

MF: C13H21NO3
MW: 239.314
CA Registry Number: [16128-88-4]; racemate [64638-06-8]
CA Chemical Name: Benzeneethanamine, 4-ethoxy-2,5-dimethoxy-α-methyl-
Category: Psychotomimetic phenethylamine
Street Names: MEM
Abuse: Limited
Type of action: Psychotomimetic, hallucinogen, euphoriant, stimulant
Human active dose: 20-50 mg
Duration of action: 10-14 hours
Toxic manifestations: Insomnia, confusion, cardiovascular disturbances, dehydration
Toxicity: Not reported

The abuse of this amphetamine has been reported in Canada [1,2]. Its preparation from a commercially available precursor (2,5-dimethoxyphenol) has also been described [3]. References 2 and 3 also describe the identification of this drug by MS, NMR and HPLC.

References:

[1] Microgram, XXI, (7), p.115, (1988);
[2] Avdovitch, et al.: Microgram, XX, (3), p.37, (1987)
[3] By, A.W. et al.: J. Forensic Sci., **35**, p.316, (1990)

Substance Number 10 in Chapter 2.1.4

IUPAC Name: 2-(6-Methoxy-benzo[1,3]dioxol-5-yl)-1-methyl-ethylamine
Synonyms: 1-(2-Methoxy-3,4-methylenedioxyphenyl)-2-propanamine
2-Methoxy-3,4-methylenedioxyamphetamine
1-Amino-2-(2-methoxy-3,4-methylenedioxyphenyl)propane
2-Methoxy-α-methyl-3,4-methylenedioxyphenethylamine
2-Methoxy-3,4-methylenedioxyphenylisopropylamine

Chemical structure:

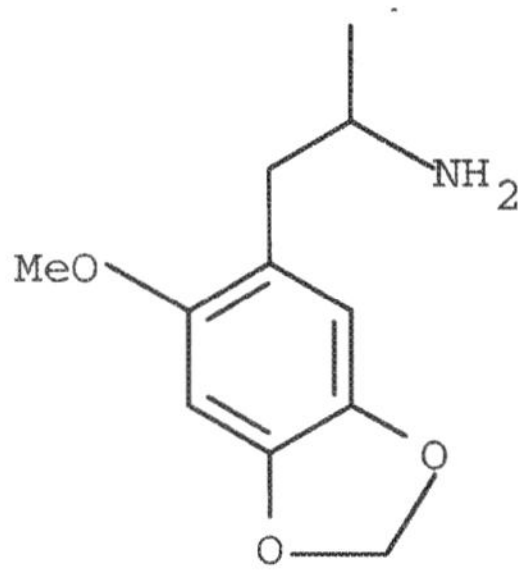

MF: C11H15N03
MW: 209.244
CA Registry Number: [23693-18-7]; racemate [64638-05-9]; racemate, hydrochloride [64778-82-1]
CA Chemical Name: 1,3-Benzodioxole-5-ethanamine, 6-methoxy-α-methyl-
Category: Psychotomimetic phenethylamine
Street Names: 2MMDA
Abuse: Limited
Type of action: Psychotomimetic, hallucinogen, euphoriant, stimulant
Human active dose: 25-50 mg
Duration of action: 8-12 hours
Toxic manifestations: Insomnia, hyperreflexia, cardiovascular disturbances, dehydration
Toxicity: Not reported

This MDA analogue was seized in a clandestine laboratory in Canada [1]. This reference also describes the identification of this compound by GC-MS and NMR.

References:

[1] Hugel, J.; Weaver, K.: Microgram, XXI, (7), p.115, (1988)

Substance Number 11 in Chapter 2.1.4

IUPAC Name: 2-(7-Methoxy-benzo[1,3]dioxol-5-yl)-1-methyl-ethylamine
Synonyms: 1-(3-Methoxy-4,5-methylenedioxyphenyl)-2-propanamine
3-Methoxy-4,5-methylenedioxyamphetamine
1-Amino-2-(3-methoxy-4,5-methylenedioxyphenyl)propane
3-Methoxy-α-methyl-4,5-methylenedioxyphenethylamine
3-Methoxy-4,5-methylenedioxyphenylisopropylamine

Chemical structure:

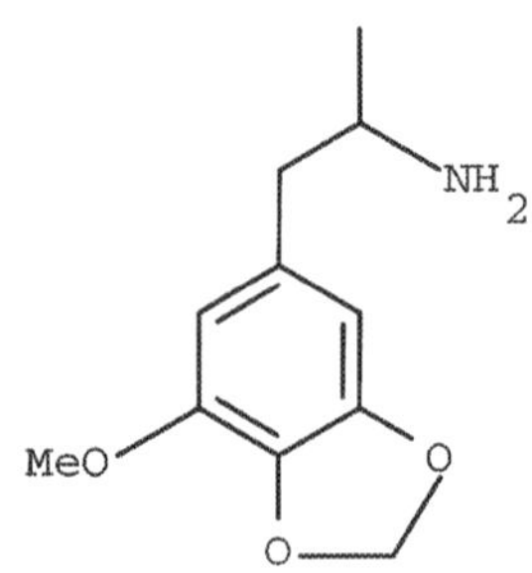

MF: C11H15NO3
MW: 209.244
CA Registry Number: [13674-05-0]; hydrochloride [33189-33-2] and [60676-84-8]; racemate [64638-02-4]
CA Chemical Name: 1,3-Benzodioxole-5-ethanamine, 7-methoxy-α-methyl-
Category: Psychotomimetic phenethylamine
Street Names: MMDA
Abuse: Limited
Type of action: Psychotomimetic, hallucinogen, euphoriant, stimulant
Human active dose: 100-250 mg
Duration of action: Moderate
Toxic manifestations: Insomnia, confusion, cardiovascular disturbances, dehydration
Toxicity: Not reported

This MDA analogue is synthesised in clandestine laboratories from myristicine, contained in readily available nutmeg oil [1]. The identification of this drug by MS and HPLC has been described [1,2].

References:

[1] Noggle, F.T.; Clark, C.R.; DeRuiter, J.: Microgram, XXVII, (10), p.321, (1995)
[2] Noggle, F.T.; Clark, C.R.; Bouhadir, K.H.; DeRuiter, J.: J. Chromatogr. Sci., **29**, p.78, (1991)

Substance Number 12 in Chapter 2.1.4

IUPAC Name: 2-(2,5-Dimethoxy-4-methylsulfanyl-phenyl)-1-methyl-ethylamine
Synonyms: 1-(2,5-Dimethoxy-4-methylthiophenyl)-2-propanamine
2,5-Dimethoxy-4-methylthioamphetamine
2,5-Dimethoxy-α-methyl-4-methylthiophenethylamine
1-Amino-2-(2,5-dimethoxy-4-methylthiophenyl)propane
2,5-Dimethoxy-4-methylthiophenylisopropylamine

Chemical structure:

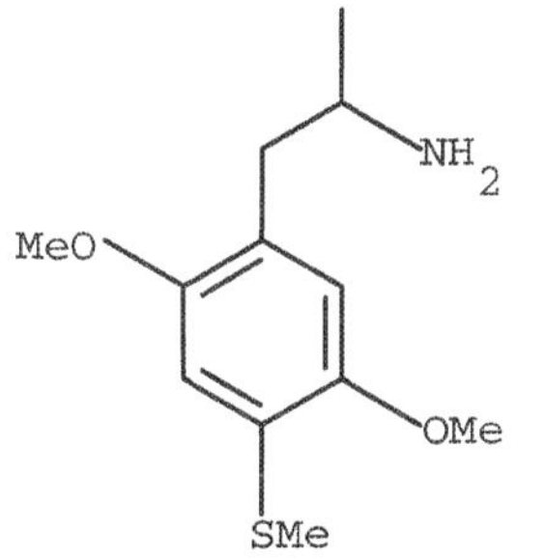

MF: C12H19NO2S
MW: 241.348
CA Registry Number: [61638-07-1]
CA Chemical Name: Benzeneethanamine, 2,5-dimethoxy-α-methyl-4-(methylthio)-
Category: Psychotomimetic phenethylamine
Street Names: Aleph1; DOT
Abuse: Rare
Type of action:, Psychotomimetic, euphoriant, hallucinogen, stimulant
Human active dose: 5-10 mg
Duration of action: 6-8 hours
Toxic manifestations: Insomnia, confusion, cardiovascular disturbances, dehydration
Toxicity: Not reported

The abuse of this amphetamine has been reported in Canada [1]. In man, this substance is a powerful euphoriant [2] with variable degrees of visual illusions and distortions. The synthesis and properties of this compound have been described [3].

References:

[1] Microgram, XXI, (7), p.114 (1988)
[2] Shulgin, A.T. and Shulgin, A.: PIHKAL, Transform Press, Inc., P.O.Box 13675, Berkeley, CA, p.463, (1991)
[3] Nichols, D.E.; Shulgin, A.T.: J. Pharm. Sci., **65**, p.1554, (1976)

Substance Number 13 in Chapter 2.1.4

IUPAC Name: 2-(4-Ethylsulfanyl-2,5-dimethoxy-phenyl)-1-methyl-ethylamine
Synonyms: 1-(4-Ethylthio-2,5-dimethoxyphenyl)-2-propanamine
4-Ethylthio-2,5-dimethoxyamphetamine
4-Ethylthio-2,5-dimethoxy-α-methylphenethylamine
1-Amino-2-(4-ethylthio-2,5-dimethoxyphenyl)propane
4-Ethylthio-2,5-dimethoxyphenylisopropylamine

Chemical structure:

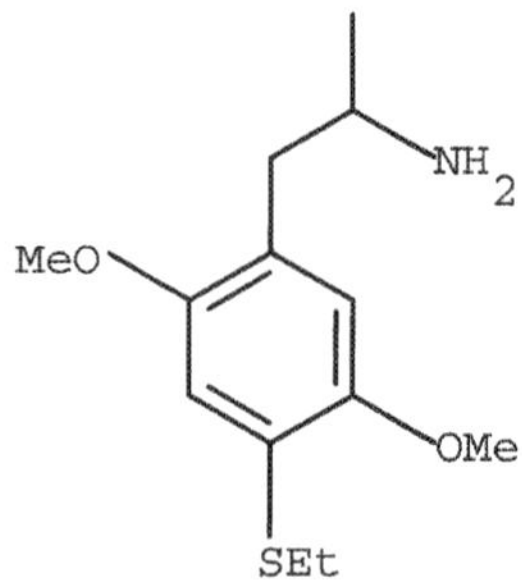

MF: C13H21NO2S
MW: 255.374
CA Registry Number: [185562-00-9]; hydrochloride [178485-02-4]
CA Chemical Name: Benzeneethanamine, 4-(ethylthio)-2,5-dimethoxy-α-methyl-
Category: Psychotomimetic phenethylamine
Street Names: Aleph2
Abuse: Rare
Type of action: Psychotomimetic, euphoriant, hallucinogen, stimulant
Human active dose: 4-8 mg
Duration of action: 8-16 hours
Toxic manifestations: Insomnia, confusion, cardiovascular disturbances, dehydration
Toxicity: Not reported

In man, this substance is a powerful psychotomimetic of a rather unpredictable nature [1].

References:

[1] Shulgin, A.T. and Shulgin, A.: PIHKAL, Transform Press, Inc., P.O.Box 13675, Berkeley, CA, p.464, (1991)

Substance Number 14 in Chapter 2.1.4

IUPAC Name: 2-(4-Ethylsulfanyl-2,5-dimethoxy-phenyl)-ethylamine
Synonyms: 1-(4-Ethylthio-2,5-dimethoxyphenyl)-2-propanamine
4-Ethylthio-2,5-dimethoxyphenethylamine

Chemical structure:

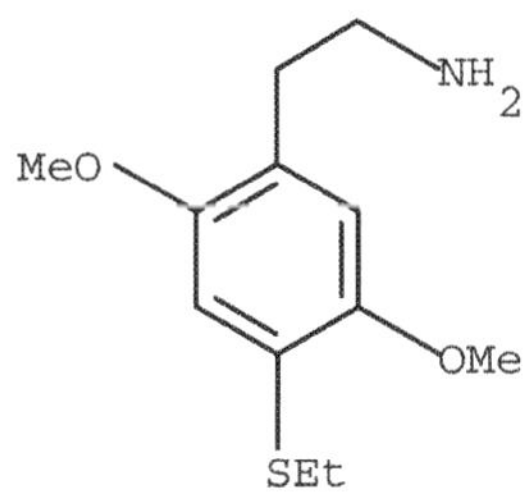

MF: C12H19NO2S
MW: 241.348
CA Registry Number: [90109-49-2]; hydrochloride [90109-50-5]
CA Chemical Name: Benzeneethanamine, 4-(ethylthio)-2,5-dimethoxy-
Category: Psychotomimetic phenethylamine
Street Names: T2
Abuse: Rare
Type of action: Psychotomimetic, hallucinogen, euphoriant
Human active dose: 12-25 mg
Duration of action: 6-8 hours
Toxic manifestations: Diarrhea, cardiovascular disturbances, dehydration
Toxicity: Not reported

Use of this phenethylamine as an aid in psychotherapy has been reported [1]. At present, the Internet monitoring shows only limited evidence of its abuse [2]. The C13-NMR spectrum of this substance has been published [3].

References:

[1] Shulgin, A.T. and Shulgin, A.: PIHKAL, Transform Press, Inc., P.O.Box 13675, Berkeley, CA, p.560, (1991)
[2] http://www.hyperreal.org/drugs/psychedelics/misc/FAQ
[3] Bienfait, B.: J. Chem. Inf. Comput. Sci., **34**, p.890, (1994)

Substance Number 15 in Chapter 2.1.4

IUPAC Name: 2-(2-Methoxy-4-methyl-5-methylsulfanyl-phenyl)-1-methyl-ethylamine
Synonyms: 1-(2-Methoxy-4-methyl-5-methylthiophenyl)-2-propanamine
2-Methoxy-4-methyl-5-methylthioamphetamine
2-Methoxy-4,α-dimethyl-5-methylthiophenethylamine
2-Methoxy-4-methyl-5-methylthiophenylisopropylamine

Chemical structure:

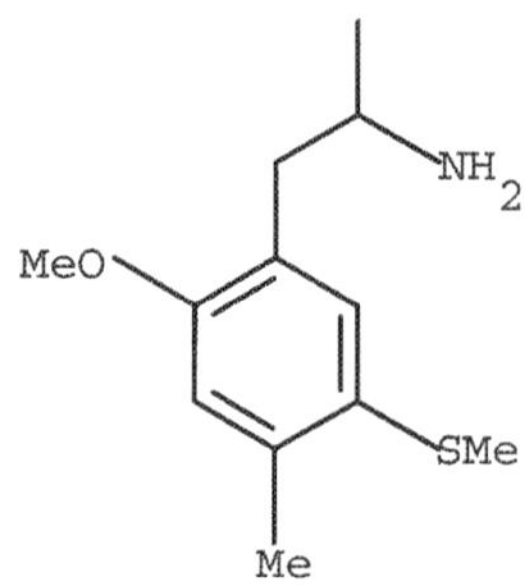

MF: C21H19NOS
MW: 225.348
CA Registry Number: Hydrochloride [84910-92-9]
CA Chemical Name: Benzeneethanamine, 2-methoxy-4,α-dimethyl-5-(methylthio)-
Category: Psychotomimetic phenethylamine
Street Names: 5TOM
Abuse: Rare
Type of action: Psychotomimetic, hallucinogen, stimulant
Human active dose: 30-50 mg
Duration of action: 6-10 hours
Toxic manifestations: Nausea, cardiovascular disturbances, dehydration
Toxicity: Not reported

A seizure of this amphetamine [1] has been reported in Canada [2].

References:

[1] Jacob, P.III.; Shulgin A.T.: J. Med. Chem., **26**, p.746, (1983)
[2] Microgram, XXI, (7), p.114, (1988)

Substance Number 16 in Chapter 2.1.4

IUPAC Name: 2-(4-Chloro-2,5-dimethoxy-phenyl)-ethylamine
Synonyms: 2-(4-Chloro-2,5-dimethoxyphenyl)ethanamine
4-Chloro-2,5-dimethoxyphenethylamine

Chemical structure:

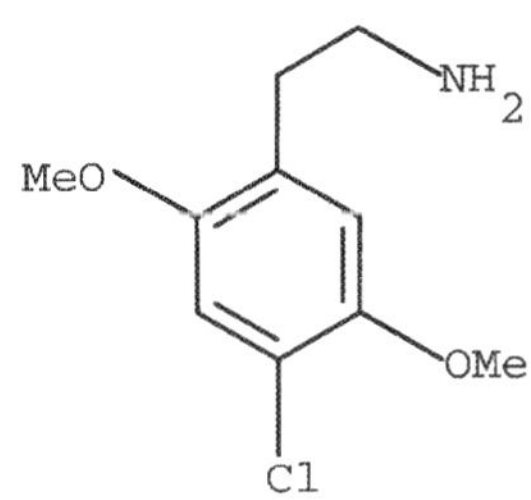

MF: $C_{10}H_{14}ClNO_2$
MW: 215.679
CA Registry Number: [88441-14-9]
CA Chemical Name: Benzeneethanamine, 4-chloro-2,5-dimethoxy-
Category: Psychotomimetic phenethylamine
Street Names: 2CC
Abuse: Rare
Type of action: Psychotomimetic, hallucinogen, aphrodisiac
Human active dose: 20-40 mg
Duration of action: 4-8 hours
Toxic manifestations: Insomnia, confusion, cardiovascular disturbances
Toxicity: Not reported

As most psychotomimetics, this drug enhances visual and auditory perception but it has a particular effect on sexual desire and pleasure similar to its bromo analogue (2CB). It also heightens all the senses such as taste and touch [1]. The preparation and properties of this compound have been reported [1,2]. The C13-NMR spectrum of this substance has also been published [3].

References:

[1] Shulgin, A.T. and Shulgin, A.: PIHKAL, Transform Press, Inc., P.O.Box 13675, Berkeley, CA, p.510, (1991)
[2] Cheng, A.C.; Castagnioli, N.: J. Med. Chem., **27**, p.513, (1984)
[3] Bienfait, B.: J. Chem. Inf. Comput. Sci., **34**, p.890, (1994)

Substance Number 17 in Chapter 2.1.4

IUPAC Name: 2-(4-Bromo-2,5-dimethoxy-phenyl)-ethylamine
Synonyms: 2-(4-Bromo-2,5-dimethoxyphenyl)ethanamine
4-Bromo-2,5-dimethoxyphenethylamine

Chemical structure:

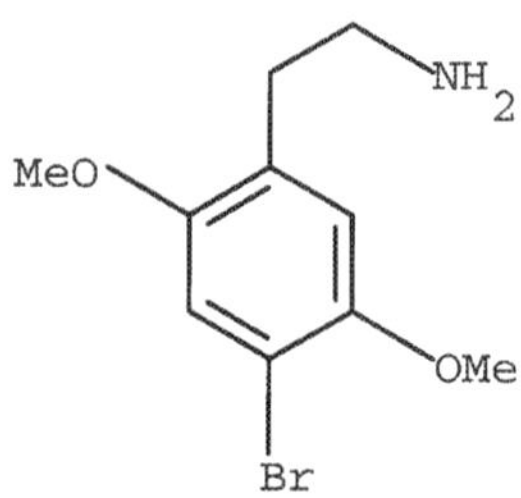

MF: C10H14BrNO2
MW: 260.130
CA Registry Number: [66142-81-2]; hydrochloride [56281-37-9]
CA Chemical Name: Benzeneethanamine, 4-bromo-2,5-dimethoxy-
Category: Psychotomimetic phenethylamine
Street Names: Venus; Nexus; Bromo; 2CB; Spectrum, BDMPEA; Toonies; MFT
Abuse: Frequent
Type of action: Psychotomimetic, hallucinogen, aphrodisiac
Human active dose: 12-24 mg
Duration of action: 4-8 hours
Toxic manifestations: Confusion, cardiovascular disturbances, dehydration
Toxicity: Doses ranging from 66 to 100 mg were accidentally absorbed giving rise to frightening intoxication which resolved rapidly without any treatment.

This drug appeared in the USA in 1979 [1]. Recently, it has become popular in Germany and Switzerland. As most psychotomimetics, this drug enhances visual and auditory perception but it has a particular effect on sexual desire and pleasure. It also heightens all the senses such as taste and touch [2].

The identification of this drug by MS and HPLC has been described [3]. The drug is usually absorbed orally in the form of capsules, tablets or dissolved in a drink, but its intranasal administration in a solution or in a powder (« snorting ») [4] is also frequent. Also, 2CB is frequently taken in a mixture with MDMA, because of the synergism of these two substances. Similarly, a dose of MDMA is often followed when its effect starts declining, by a dose of 2CB to obtain a new desirable effect [5]. In low dosage range (about 10 mg) the drug is popular as an « aesthetic enhancer » when going to concerts or galleries.

References:

[1] Fed. Reg., **60**, No.106, p.28718, (1995); Microgram, XXVIII, (8), p.242, (1995)
[2] Shulgin, A.T.; Carter, M.F.: Psychopharmacology Commun., **1**, p.93, (1975)
[3] DeRuiter, J.; Clark, C.R.; Noggle, F.T.: J. Chromatogr. Sci., **33**, p.583, (1995)
[4] http://www.damicon.fi/drugs/psychedelics/phenethylamines/2cb.info
[5] http:// www.ecstasy.org/combinations.html

Substance Number 18 in Chapter 2.1.4

IUPAC Name: 2-(4-Iodo-2,5-dimethoxy-phenyl)-ethylamine
Synonyms: 2-(4-Iodo-2,5-dimethoxyphenyl)ethanamine
4-Iodo-2,5-dimethoxyphenethylamine

Chemical structure:

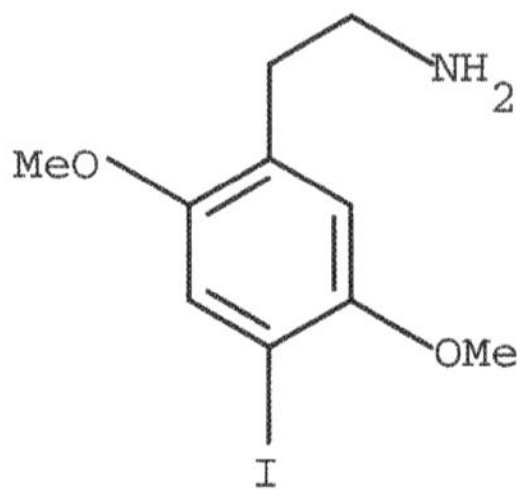

MF: C10H14 NOI
MW: 307.131
CA Registry Number: [69587-11-7]
CA Chemical Name: Benzeneethanamine, 4-iodo-2,5-dimethoxy-
Category: Psychotomimetic phenethylamine
Street Names: 2CI
Abuse: Rare
Type of action: Psychotomimetic, hallucinogen, aphrodisiac
Human active dose: 14-22 mg
Duration of action: 6-10 hours
Toxic manifestations: Confusion, cardiovascular disturbances, dehydration
Toxicity: Not reported

As most psychotomimetics this drug enhances visual and auditory perception in man, but it has a particular effect on sexual desire and pleasure. In this respect, its effects strongly resemble 2CB. It also heightens all the senses such as taste and touch [1].

References:

[1] Shulgin, A.T. and Shulgin, A.: PIHKAL, Transform Press, Inc., P.O.Box 13675, Berkeley, CA, p.540, (1991)

2.2 LSD Analogues

Being extremely potent and relatively easy to produce, LSD (**1**, R1=R2=Et;) is a drug particularly suited for illicit traffic. Several procedures for its manufacture have been developed and mastered by clandestine chemists [1]. However, diethylamine, the key precursor in LSD synthesis is a highly suspect and closely controlled chemical and has been occasionally replaced by another suitable dialkylamine [2]. Consequently, in spite of their generally lower potency, several LSD analogues have sporadically been detected on the drug market.

Thus, the ethylpropylamide (LEP-57; **1**, R1=Et; R2=n-Pr) shows about 50% of LSD psychotomimetic activity in man. The morpholide (**2**) and the methylpropylamide (LMP-55, LAMPA; **1**, R1=Me; R2=n-Pr) [7] are still less active (30% [3], 10-15 % [4,5], respectively). The identification of these substances was described in several recent papers [6].

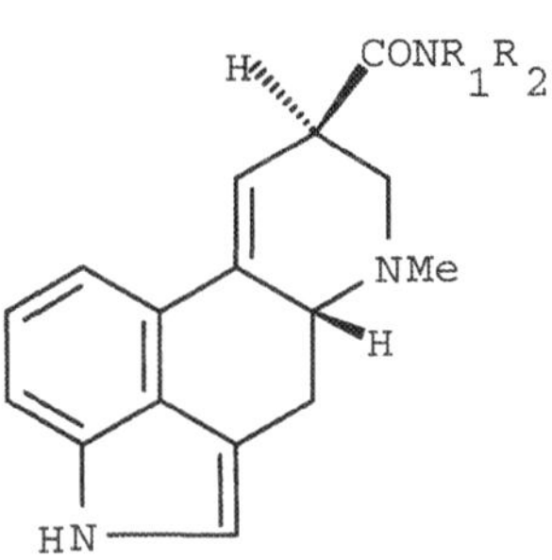

1

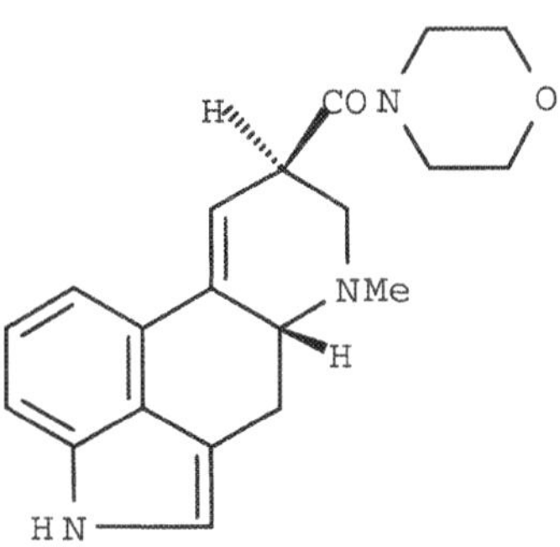

2

0.15-0.4 mg

Most of the other LSD analogues (e.g. piperidide, pyrrolidide) show very low or negligible LSD-like activity in humans [6]. Only two LSD analogues have recently been reported to be more potent psychotomimetics than LSD itself [8]. However, there is no evidence of their eventual abuse or clandestine manufacture.

Out of the LSD analogues mentioned above, the lysergic acid morpholide (LSM) (**2**) may be of particular interest to the clandestine chemist because of its very easy preparation.

2.2.1 General References

[1] Uncle Fester: Practical LSD manufacture, Loompanics Unlimited, Port Townsend, WA 98368, (1995)

[2] Smith, M.V.: Psychedelic Chemistry, Loompanics Unlimited, Box 1197, Port Townsend, WA 98368, pp.103-137, (1992)

[3] Gogerty, J.H. et al.: J. Pharmacol., **120**, p.340, (1957)

[4] LSD-Total Study, D.V. Siva Sankar, Edit., PJD Publications Ltd., Westbury, NY 11590; p.65, (1975)
[5] Kumbar, M. et al.: Res. Comm. Chem. Pathol. Pharmacol., **6**, p.65, (1973)
[6] Japp, M. et al.: J. Forensic Sci., **32**, p.933, (1987)
[7] Mills III, T.; Roberson, J.C.; McCurdy, H.H.; Wall, W.H.: Instrumental Data for Drug Analysis, Elsevier Science Publishing Co., Inc., p.1292, (1992),
[8] Hoffmann, A.J.; Nichols, D.E.: Psychopharmacology, **91**, p.67, (1987)

2.2.2 Data Sheets : LSD Analogues

Substance Number 2 in Chapter 2.2

IUPAC Name: (7-Methyl-4,6,6a,7,8,9-hexahydro-indolo[4.3-f,g]quinolin-9-yl)-morpholin-4-yl-methanone
Synonyms: 4-[(8β)-6-Methylergol-9-ene-8-yl)carbonyl]-1,4-oxazine
D-Lysergic acid morpholide

Chemical structure:

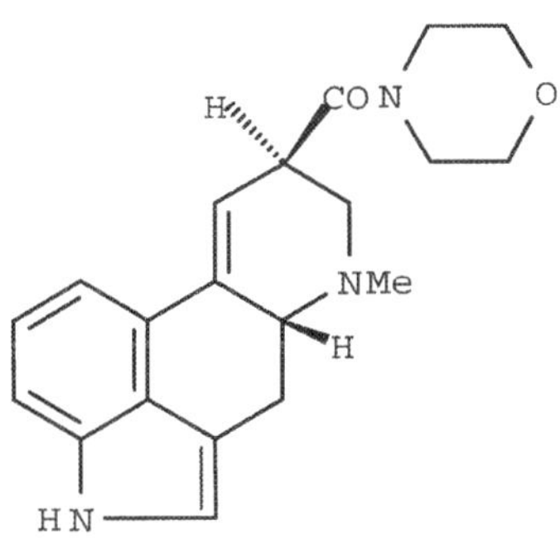

MF: C20H23N3O2
MW: 337.421
CA Registry Number: [4314-63-0]
CA Chemical Name: Morpholine, 4-[(8β)-9,10-didehydro-6-methylergolin-8-yl)carbonyl]-
Category: Psychotomimetic LSD analogue
Street Names: LSM
Abuse: Rare
Type of action: Psychotomimetic, hallucinogen
Human active dose: 0.15-0.4 mg
Duration of action: 5-8 hours
Toxic manifestations: Tachycardie, hypertension, mydriasis, anorexia, ataxia, hyperreflexia, respiratory depression, coma
Toxicity: LD50 is 0.6 mg/kg i.v. (rabbit). In man, the LD 50 is estimated to about 0.2 mg/kg p.o.[1].

In man, this substance shows practically the same effects as LSD [1]. Among the psychotomimetic LSD analogues, this compound may be of particular interest to a clandestine manufacturer because it can be easily prepared by simple heating of ergotamine in morpholine [2].

References:

[1] LSD-Total Study, D.V. Siva Sankar, Edit., PJD Publications Ltd. Westbury, NY 11590; p.65, (1975)
[2] Valter, K.: Unpublished case report, (1980)

2.3 Psychotomimetic Indolealkylamines

The indole derivatives included in this chapter produce, in man, psychic effects very similar to those of LSD. Derealisation, depersonalisation, visual illusions, hallucinations, altered colours, sounds, space perceptions and synesthesias are consistently present. Physiologic effects such as tachycardia, marked mydriasis, ataxia, tremor, hyperreflexia, nausea, hypertension and occasionally hyperthermia are also observed. Clinical management of poisoning by this type of drug has been thoroughly described [1].

Before becoming known as designer drugs, several substances in this chapter were already found in nature [2].

Thus, N,N-dimethyltryptamine (DMT, **1**), 5-methoxy-N,N-dimethyl-tryptamine (5MEODMT, **2**), 4-hydroxy-N,N-dimethyltryptamine (psilocine, **3**) and psilocybine (O-phosphorylpsilocine) were found in a large number of plants, mushrooms and even in animals [2]. The two latter very important drugs of abuse, coming almost exclusively from natural sources, are not considered as designer drugs.

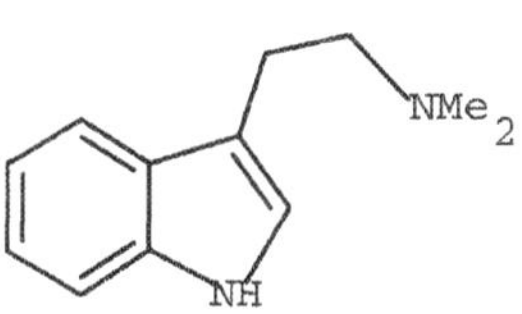

1

50-70 mg

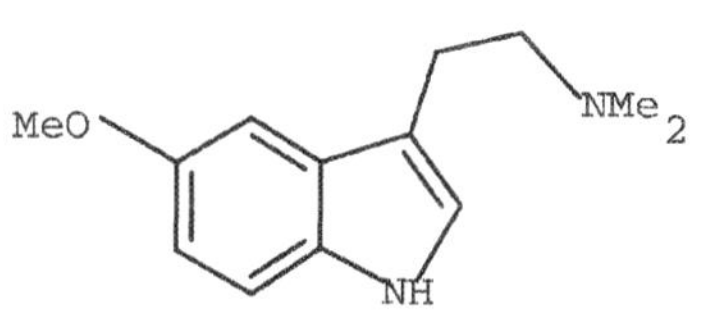

2

4-8 mg

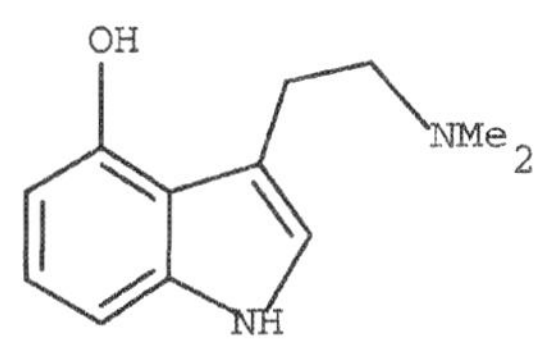

3

8-15 mg

DMT and 5MEODMT are practically inactive when administered orally [3]. Therefore, these drugs are injected or smoked. Various smokable preparations, particularly parsley or cannabis leaves impregnated with the drugs, have been described [4].

Interestingly, on inhalation of these drugs, only about a third of the dose indicated above is needed for a full psychotomimetic effect with very rapid onset. The experience of very short duration has been described as very impressive [5,6].

Chemical and structural modifications of the compounds above have given rise to a number of potent, orally active analogues such as DET (**4**), DPT (**5**) [6], DIPT (**6**) and DAT (**7**) [7]. The last two compounds are the most potent psychotomimetics among unsubstituted N,N-dialkyltryptamines.

Recently, new tryptamines of high abuse potential (**8**, **9**) have been synthesized [8]. Thus, the psilocine derivative (**9**) exhibits in man a very intense and particular psychotomimetic activity.

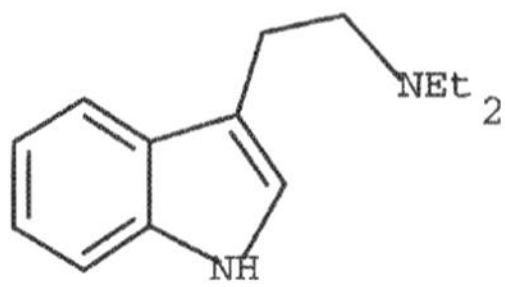

4

50-70 mg

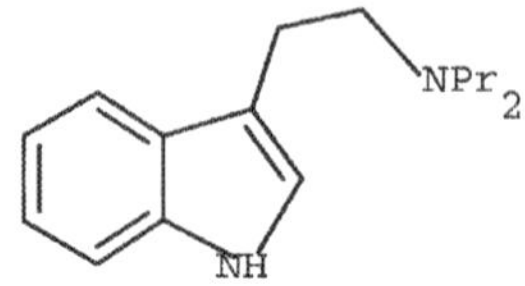

5

50-70 mg

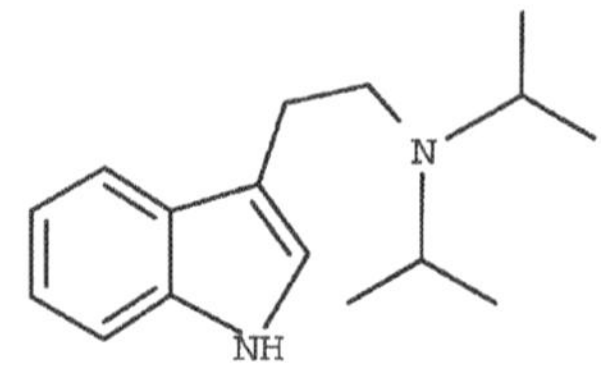

6

25-35 mg

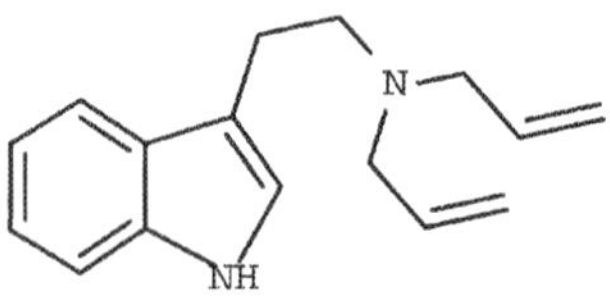

7

30-50 mg

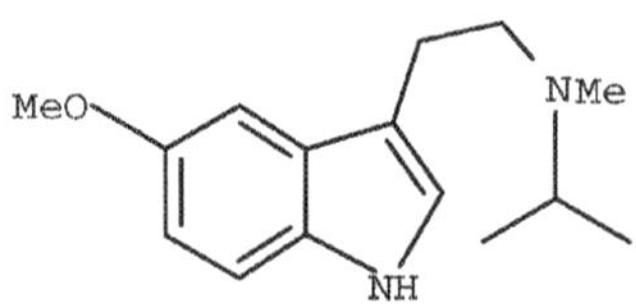

8

3-5 mg

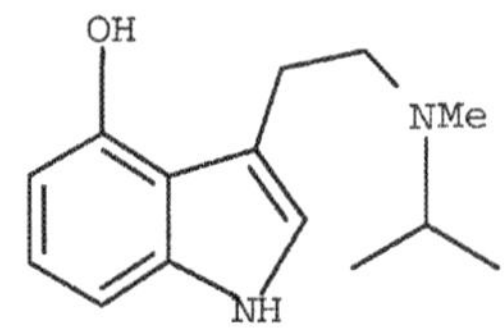

9

10 mg

Another powerful psychotomimetics were obtained by simple lateral chain modification (**10, 11, 12**) [9,10].

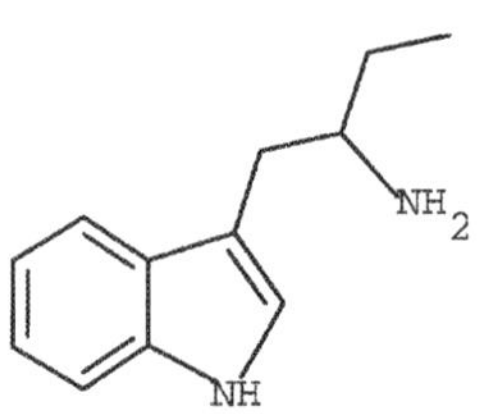

10

100-150 mg

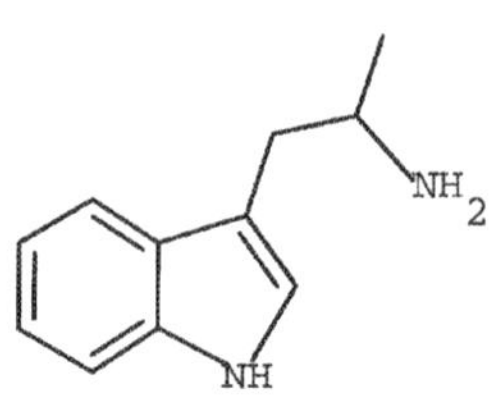

11

15-35 mg

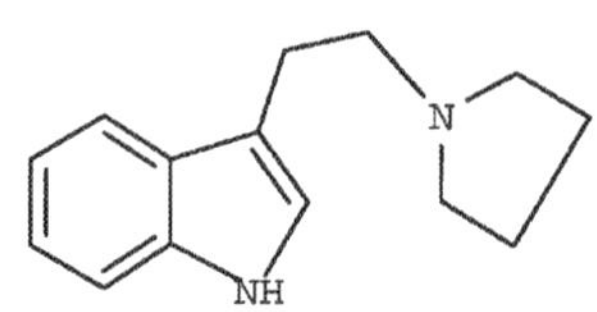

12

50-70 mg

For a moment, the abuse of these last substances is only sporadic.

Only the etryptamine (ET, **10**) appears frequently at « Rave » parties. It seems to defy all pharmacological classification but it has been reported to possess the MDMA-like entactogenic properties.

2.3.1 General References

[1] Litovitz, T.: Hallucinogens. In: Clinical Management of Poisoning and Drug Overdose, Eds. Haddad, L.M.; Winchester, J.F.; W.B. Saunders Company, p.455, (1983)

[2] Schultes, R.E.; Hofmann, A.: Plants of the Gods, Healing Arts Press, One Park Street, Rochester, VT 05767, pp.40 and 120, (1992)

[3] Stafford, P.: Psychedelic Encyclopedia, Ronin Publishing, Inc., Box 1035, Berkeley, CA 94701, p.392, (1992)
[4] ibid., p.322
[5] ibid., p.322
[6] ibid., p.319
[7] ibid., p.320
[8] Repke, D.B.; Grotjahn, D.B.; Shulgin, A.T.: J. Med. Chem., **28**, p.892, (1985)
[9] Kantor, R.E. et al.: Biological Psychiatry, **15**, p.349, (1980)
[10] Shulgin, A.T.: Chemistry of Psychotomimetics. In: Hoffmeister, F. & Stille, G. (Eds.) Handbook of Experimental Pharmacology, Vol. **55**: Alcohol and Psychotomimetics, Psychotropic Effects of Central-Acting Drugs, Springer-Verlag, New York, (1982)

2.3.2 Data Sheets: Psychotomimetic Indolealkylamines

Substance Number 1 in Chapter 2.3

IUPAC Name: [2-(1H-Indol-3-yl)-ethyl]-dimethyl-amine
Synonyms: 3-(2-Dimethylaminoethyl)indole
N,N-Dimethyltryptamine

Chemical Structure:

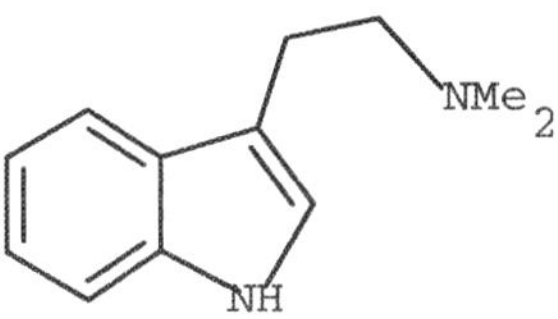

MF: C12H16N2
MW: 188.272
CA Registry Number: [61-50-7]
CA Chemical Name: 1H-Indole-3-ethanamine, N,N-dimethyl-
Category: Psychotomimetic indolealkylamine
Street Names: DMT
Abuse: Frequent
Type of action: Psychotomimetic, hallucinogen
Human active dose: 50-70 mg (injected); 15-20 mg (inhaled)
Duration of action: 15-30 min.
Toxic manifestations: Tachycardia, hypertension, mydriasis, anorexia, ataxia, hyperreflexia, hyperthermia, coma
Toxicity: Not reported

This tryptamine shows a very short but strong hallucinogenic action in man [1]. This substance must be inhaled (smoked) or injected to show any activity [1,2]. In fact, oral doses up to 1000 mg have been administered to humans without any noticeable effect. A page devoted to this drug may be found on the Internet [2]. The drug found on the clandestine market [3] is only partly produced by synthesis because it is also currently obtained by extraction of plants (particularly of Phalaris sp.) [4]. Due to its simplicity and absence of any control, the extraction has usually been preferred to the synthesis but the situation may change in the future. Actually, the recently described new single step synthesis of various dialkyltryptamines [5], which is well suited to DMT manufacture, is familiar to clandestine chemists. The identification of this drug by MS and HPLC has been described [6].

References:

[1] Shulgin, A.T.: Profiles of psychedelic drugs. 1. DMT. J. of Psychedelic Drugs, **8**, p.167, (1976)
[2] http://hyperreal.com/drugs/psychedelics/dmt.htm
[3] Microgram, XVII, (3), p.33, (1984)
[4] Smith, T.A.: Phytochemistry, **16**, p.171, (1977)
[5] Cheng-yi, C.; Senanayake, C.H.; Bill, T.J.; Larsen, R.D.; Verhoeven, T.R.; Reider, P.J.: J. Org. Chem., **59**, p.3738, (1994)
[6] Mills III, T.; Roberson, J.C.; McCurdy, H.H.; Wall, W.H.: Instrumental Data for Drug Analysis, Elsevier Science Publishing Co., Inc., p.742, (1992)

Substance Number 2 in Chapter 2.3

IUPAC Name: [2-(5-Methoxy-1H-Indol-3-yl)-ethyl]-dimethyl-amine
Synonyms: 3-(2-Dimethylaminoethyl)-5-methoxyindole
5-Methoxy-N,N-dimethyltryptamine

Chemical Structure:

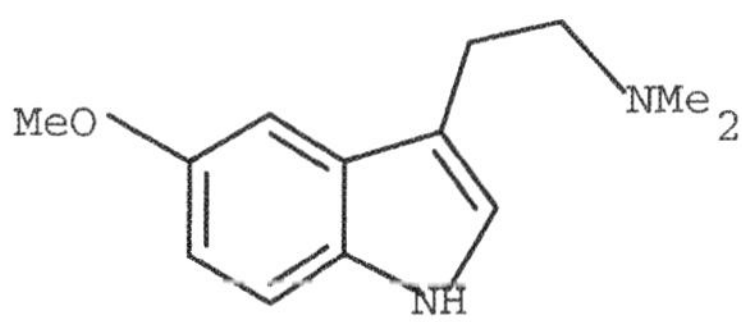

MF: C13H18N2O
MW: 218.298
CA Registry Number: [1019-45-0]; hydrochloride [2427-79-4]
CA Chemical Name: 1H-Indole-3-ethanamine, 5-methoxy-N,N-dimethyl-
Category: Psychotomimetic indolealkylamine
Street Names: 5MEODMT
Abuse: Limited
Type of action: Psychotomimetic, hallucinogen
Human active dose: 4-8 mg
Duration of action: 1 hour
Toxic manifestations: Tachycardia, hypertension, mydriasis, anorexia, ataxia, hyperreflexia, hyperthermia, coma
Toxicity: Not reported

This drug of abuse [1] is frequently smoked because its oral absorption tends to be irregular. In man, it shows strong hallucinogenic activity (sometimes described as « overwhelming »), which is similar to that of N,N-dimethyltryptamine (DMT). The substance is obtained by synthesis or extracted from natural sources, including animals. Actually, this substance is contained (in concentration up to 15%) in the venom of a small Mexican toad found in Sonora desert [2]. In this last case, the dried venom can be directly smoked. The synthesis of this substance is likely to become more frequent in the future. In fact, the recently described new single step synthesis of various dialkyltryptamines [3], starting from commercially available precursors, represents easy access to this particular tryptamine. The identification of this drug by MS and HPLC has been described [4].

References:

[1] Microgram, XXIV, (9), p.218, (1991)
[2] Chamakura, R.P.: Microgram, XXVII, (9), p.319, (1994)
[3] Cheng-yi, C.; Senanayake, C.H.; Bill, T.J.; Larsen, R.D.; Verhoeven, T.R.; Reider, P.J.: J. Org. Chem., **59**, p.3738, (1994)
[4] Mills III, T.; Roberson, J.C.; McCurdy, H.H.; Wall, W.H.: Instrumental Data for Drug Analysis, Elsevier Science Publishing Co., Inc., p.1422, (1992)

Substance Number 4 in Chapter 2.3

IUPAC Name: Diethyl-[2-(1H-indol-3-yl)-ethyl]-amine
Synonyms: 3-(2-Diethylaminoethyl)indole
N,N-Diethyltryptamine

Chemical Structure:

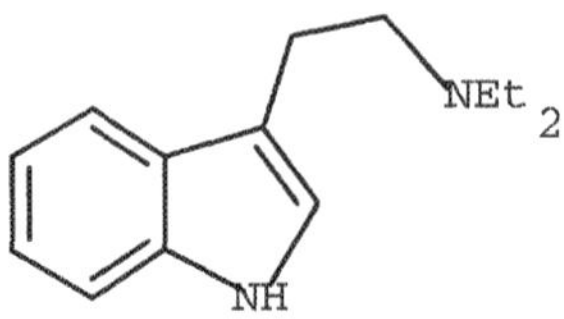

MF: C14H20N2
MW: 216.325
CA Registry Number: [61-51-8]
CA Chemical Name: 1H-Indole-3-ethanamine, N,N-diethyl-
Category: Psychotomimetic indolealkylamine
Street Names: DET
Abuse: Rare
Type of action: Psychotomimetic
Human active dose: 50-70 mg
Duration of action: 3-5 hours
Toxic manifestations: Tachycardia, hypertension, mydriasis, anorexia, ataxia, hyperreflexia, hyperthermia, coma
Toxicity: Not reported

This tryptamine is not found in natural sources. When injected or inhaled, it shows strong hallucinogenic activity, similar to that of LSD [1].
The synthesis of this substance is well mastered by clandestine operators [2]. The recently described synthesis of various dialkyltryptamines [3], starting from commercially available precursors, represents an easy way to this tryptamine. The identification of this substance by MS and HPLC has been described [4].

References:

[1] Szara, S.; Rockland, L.H.; Rosenthal, D.; Handlon, J.H.: Arch. Gen. Psychiat., **15**, p.321, (1966)
[2] Smith, M.V.: Psychedelic Chemistry, Loompanics Unlimited, Box 1197, Port Townsend, WA 98368, p.40, (1981)
[3] Cheng-yi, C.; Senanayake, C.H.; Bill, T.J.; Larsen, R.D.; Verhoeven, T.R.; Reider, P.J.: J. Org. Chem., **59**, p.3738, (1994)
[4] Mills III, T.; Roberson, J.C.; McCurdy, H.H.; Wall, W.H.: Instrumental Data for Drug Analysis, Elsevier Science Publishing Co., Inc., p.670, (1992)

Substance Number 5 in Chapter 2.3

IUPAC Name: [2-(1H-Indol-3-yl)-ethyl]-dipropyl-amine
Synonyms: 3-(2-Dipropylaminoethyl)indole
N,N-Dipropyltryptamine

Chemical Structure:

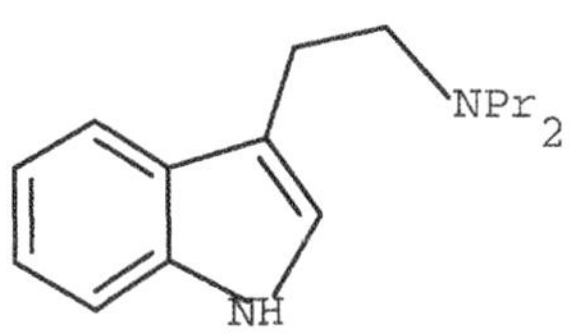

MF: C16H24N2
MW: 244.379
CA Registry Number: [61-52-9]
CA Chemical Name: 1H-Indole-3-ethanamine, N,N-dipropyl-
Category: Psychotomimetic indolealkylamine
Street Names: DPT
Abuse: Rare
Type of action: Psychotomimetic, hallucinogen
Human active dose: 50-70 mg
Duration of action: 4-6 hours
Toxic manifestations: Tachycardia, hypertension, mydriasis, anorexia, ataxia, hyperreflexia, hyperthermia, coma
Toxicity: Not reported

This orally active tryptamine is not found in natural sources. Use of this drug in psychotherapy has been reported [1].

The synthesis of this substance is well mastered by clandestine operators [2,4]. The recently described new single step synthesis of various dialkyltryptamines [3], starting from commercially available precursors, represents an easy way to this tryptamine. The identification of this substance by MS and HPLC has been described [5].

References:

[1] Grof, S.; Richards, W.A.: Arch. Gen. Psychiatry, **28**, p.817, (1973)
[2] Microgram, XXI, (8), p.114, (1988)
[3] Cheng-yi, C.; Senanayake, C.H.; Bill, T.J.; Larsen, R.D.; Verhoeven, T.R.; Reider, P.J.: J. Org. Chem., **59**, p.3738, (1994)
[4] Smith, M.V.: Psychedelic Chemistry, Loompanics Unlimited, Box 1197, Port Townsend, WA 98368, p.40, (1981)
[5] Mills III, T.; Roberson, J.C.; McCurdy, H.H.; Wall, W.H.: Instrumental Data for Drug Analysis, Elsevier Science Publishing Co., Inc., p.670, (1992)

Substance Number 6 in Chapter 2.3

IUPAC Name: [2-(1H-Indol-3-yl)-ethyl]-diisopropyl-amine
Synonyms: 3-(2-Diisopropylaminoethyl)indole
N,N-Diisopropyltryptamine

Chemical Structure:

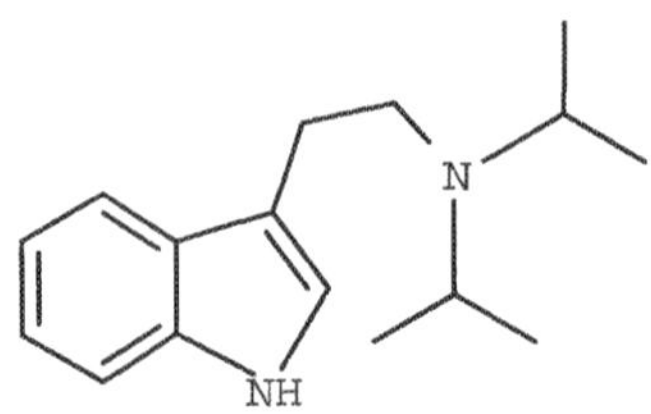

MF: C16H24N2
MW: 244.379
CA Registry Number: [14780-24-6]
CA Chemical Name: 1H-Indole-3-ethanamine, N,N-bis(1-methylethyl)-
Category: Psychotomimetic indolealkylamine
Street Names: DIPT
Abuse: Rare
Type of action: Psychotomimetic, hallucinogen
Human active dose: 25-35 mg
Duration of action: 4-6 hours
Toxic manifestations: Tachycardia, hypertension, mydriasis, anorexia, ataxia, hyperreflexia, hyperthermia, coma
Toxicity: Not reported

This orally active tryptamine [1] is not found in natural sources. The synthesis of this substance is well mastered by clandestine operators [2]. The recently described synthesis of various dialkyltryptamines [3], starting from commercially available precursors, is well suited to the preparation of this particular tryptamine.

References:

[1] Shulgin, A.T.; Carter, M.F.: Commun. Psychopharmacol., **4**, p.363, (1981)
[2] Smith, M.V.: Psychedelic Chemistry, Loompanics Unlimited, Box 1197, Port Townsend, WA 98368, p.40, (1981)
[3] Cheng-yi, C.; Senanayake, C.H.; Bill, T.J.; Larsen, R.D.; Verhoeven, T.R.; Reider, P.J.: J. Org. Chem., **59**, p.3738, (1994)

Substance Number 7 in Chapter 2.3

IUPAC Name: Diallyl-[2-(1H-indol-3-yl)-ethyl]-amine
Synonyms: 3-(2-Diallylaminoethyl)indole
N,N-Diallyltryptamine

Chemical Structure:

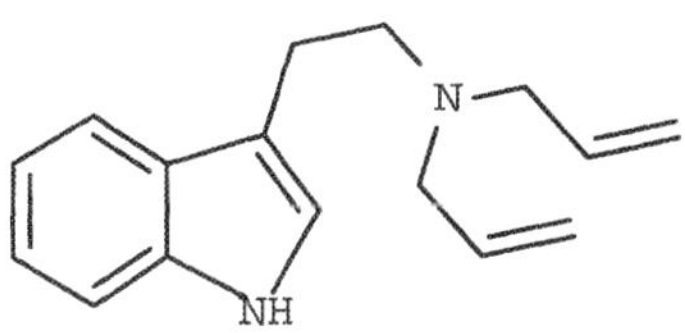

MF: C16H20N2
MW: 240.347
CA Registry Number: [60676-77-9]
CA Chemical Name: 1H-Indole-3-ethanamine, N,N-di-2-propenyl-
Category: Psychotomimetic indolealkylamine
Street Names: DAT
Abuse: Uncommon
Type of action: Psychotomimetic, hallucinogen
Human active dose: 30-50 mg
Duration of action: 4-6 hours
Toxic manifestations: Tachycardia, hypertension, mydriasis, anorexia, ataxia, hyperreflexia, hyperthermia, coma
Toxicity: Not reported

This orally active tryptamine [1,2] is not found in natural sources. The recently described new single step synthesis of various dialkyltryptamines [3], starting from commercially available precursors, is well suited to the preparation of this particular tryptamine.

References:

[1] Tryptamine Carriers.: http://www.levity.com/deoxy/ trypfaq.htm (Last update: Dec 1995 by P. Pennanen)
[2] Szara, S.: DMT (N,N-dimethyltryptamine) and homologues, clinical and pharmacological considerations. In: Psychotomimetic drugs, Efron. D. Ed. Raven Press, New York, p.275, (1970)
[3] Cheng-yi, C.; Senanayake, C.H.; Bill, T.J.; Larsen, R.D.; Verhoeven, T.R.; Reider, P.J.: J. Org. Chem., **59**, p.3738, (1994)

Substance Number 8 in Chapter 2.3

IUPAC Name: Isopropyl-[2-(5-methoxy-1H-indol-3-yl)-ethyl]-methyl-amine
Synonyms: 3-[2-(N-Isopropyl-N-methylamino)ethyl]-5-methoxyindole N-Isopropyl-N-methyl-5-methoxytryptamine

Chemical Structure:

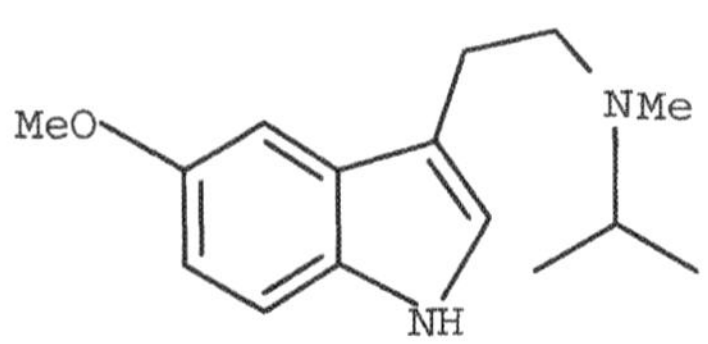

MF: C15H22N2O
MW: 246.352
CA Registry Number: [96096-55-8]
CA Chemical Name: 1H-Indole-3-ethanamine, 5-methoxy-N-methyl-N-(1-methylethyl)-
Category: Psychotomimetic indolealkylamine
Street Names: Not reported
Abuse: Rare
Type of action: Stimulant, psychotomimetic
Human active dose: 3-5 mg
Duration of action: 3 hours
Toxic manifestations: Tachycardia, hypertension, mydriasis, anorexia, ataxia, hyperreflexia, hyperthermia, coma
Toxicity: Not reported

In man, this potent, orally active tryptamine [1] shows CNS stimulant and psychotomimetic properties. At present, its abuse is limited to isolated cases. The recently described synthesis of various dialkyltryptamines [2], starting from commercially available precursors, is also suited to the preparation of this particular tryptamine.

References:

[1] Repke, D. B.; Grotjahn, D. B.; Shulgin, A. T.: J. Med. Chem., **28**, p.892, (1985)
[2] Cheng-yi, C.; Senanayake, C.H.; Bill, T.J.; Larsen, R.D.; Verhoeven, T.R.; Reider, P.J.: J. Org. Chem., **59**, p.3738, (1994)

Substance Number 9 in Chapter 2.3

IUPAC Name: 3-[2-(Isopropyl-methyl-amino)-ethyl]-1H-indol-4-ol
Synonyms: 4-Hydroxy-3-[2-(N-isopropyl-N-methylamino)ethyl]indole
4-Hydroxy-N-isopropyl-N-methyltryptamine

Chemical Structure:

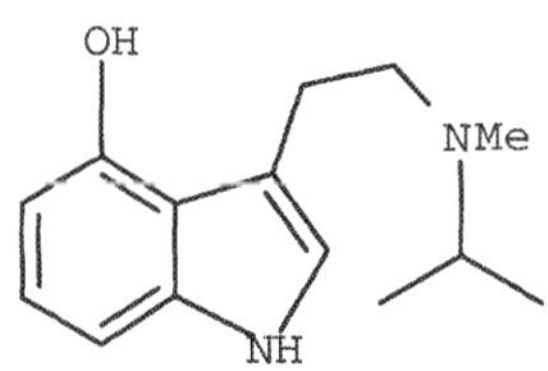

MF: C14H20N2O
MW: 232.325
CA Registry Number: [77872-43-6]
CA Chemical Name: 1H-Indol-4-ol, 3-[[2-(methyl(1- methylethyl) amino]ethyl]-
Category: Psychotomimetic indolealkylamine
Street Names: Not reported
Abuse: Uncommon
Type of action: Psychotomimetic, hallucinogen
Human active dose: 10 mg
Duration of action: 6-7 hours
Toxic manifestations: Tachycardia, hypertension, mydriasis, anorexia, ataxia, hyperreflexia, hyperthermia, coma
Toxicity: Not reported

This drug of high abuse potential produces very intense hallucinations in man [1,2].

References:

[1] Repke, D.B.; Grotjahn, D.B.; Shulgin, A.T.: J. Med. Chem., **28**, p.892, (1985)
[2] Tryptamine Carriers.: http://www.levity.com/deoxy/ trypfaq.htm (Last update: Dec. 1995 by P. Pennanen)

Substance Number 10 in Chapter 2.3

IUPAC Name: 1-(1H-Indol-3-yl-methyl)-propylamine
Synonyms: 3-[1-(2-Aminobutyl]indole
α-Ethyltryptamine
Etryptamine

Chemical Structure:

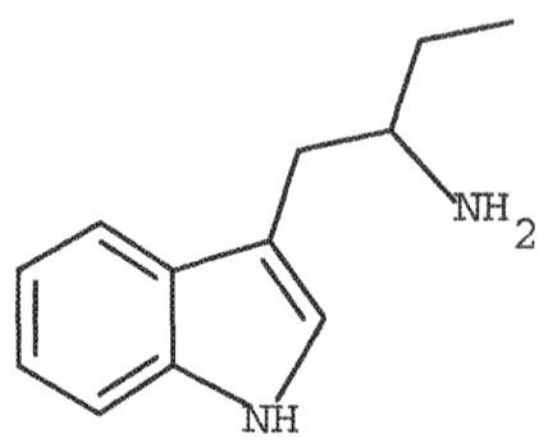

MF: C11H14N2
MW: 187.264
CA Registry Number: [2235-90-7]
CA Chemical Name: 1H-Indol-3-ethanamine, α-ethyl-
Category: Psychotomimetic indolealkylamine
Street Names: ET; TRIP
Abuse: Frequent
Type of action: Entactogen, euphoriant, psychotomimetic
Human active dose: 100-150 mg (as acetate)
Duration of action: Long
Toxic manifestations: Mydriasis, anorexia, ataxia, hyperreflexia, hepatotoxicity
Toxicity: LD50 is 48 mg/kg, p.o. (mouse). Estimated 700 mg dose has been involved in a fatal intoxication in Germany [4].

This tryptamine was reported to have MDMA-like, entactogenic effects in man [1] and be abused at « Rave parties » in Germany [4]. Its clandestine manufacture and several cases of lethal intoxication involving this drug have been reported [2,3,4]. This substance used to be marketed as an antidepressant (« Monase ») of the IMAO type [3], but commercial distribution had to be discontinued due to drug hepatotoxicity. The identification of this drug by MS and HPLC has been described [4].

References:

[1] Glennon, R.A.: Pharmacol. Biochem. Behav., **46**, p.459, (1993)
[2] Morano, R.A.; Spies, C.; Walker, F.B.; Plank, S.M.: J. Forensic Sci., **38**, p.721, (1993)
[3] Microgram, XXVII, (11), p.367, (1994)
[4] Daldrup, T.; Heller, C.; Matthiesen, U.; Honus, S.; Bresges, A.; Haarhoff, K.: Z. Rechtsmed., **97**, p.61, (1986)

Substance Number 11 in Chapter 2.3

IUPAC Name: 2-(1H-Indol-3-yl)-1-methyl-ethylamine
Synonyms: 3-[1-(2-Aminopropyl)]indole
α-Methyltryptamine

Chemical Structure:

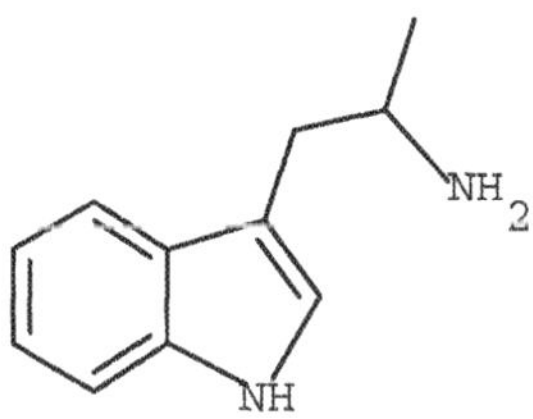

MF: C11H14N2O
MW: 190.244
CA Registry Number: [299-26-3]; racemate [304-54-1]; [S]-7795-51-9]; [R]-7795-52-0]; hydrochloride [879-36-7]
CA Chemical Name: 1H-Indol-3-ethanamine, α-methyl-
Category: Psychotomimetic indolealkylamine
Street Names: AMT; Rolls-Royce
Abuse: Rare
Type of action: Psychotomimetic
Human active dose: 15-35 mg
Duration of action: More than 12 hours
Toxic manifestations: Tachycardia, hypertension, mydriasis, anorexia, ataxia, hyperreflexia, hyperthermia, coma
Toxicity: LD50 is 22.4 mg/kg, p.o. (mouse).

In man, this potent orally active tryptamine shows clear hallucinogenic properties [1]. Interestingly, this substance used to be employed as an antidepressant, particularly in Russia (« Indopan »).

At present, its abuse is limited to isolated cases [2]. The identification of this drug by GC-MS [3] and NMR [4] has been described.

References:

[1] Murphree, H.B.; Dippy, R.H.; Jenney, E.H.; Pfeiffer, C.C.: Clin. Pharmacol. Ther., **2**, p.722, (1961)
[2] Smith, M.V.: Psychedelic Chemistry, Loompanics Unlimited, Box 1197, Port Townsend, WA 98368, p.40, (1981)
[3] Narasimhachari, N.; Leiner, K.: J. Chromatogr. Sci., **15**, p.181, (1977)
[4] Repke, D.B.; Ferguson, W.J.: J. Heterocycl. Chem., **13**, p.775, (1976)

Substance Number 12 in Chapter 2.3

IUPAC Name: 3-(2-Pyrrolidin-1-yl-ethyl)-1H-indole
Synonyms: 3-[2-(1-Pyrrolidyl)ethyl]indole
N,N-Tetramethylenetryptamine

Chemical Structure:

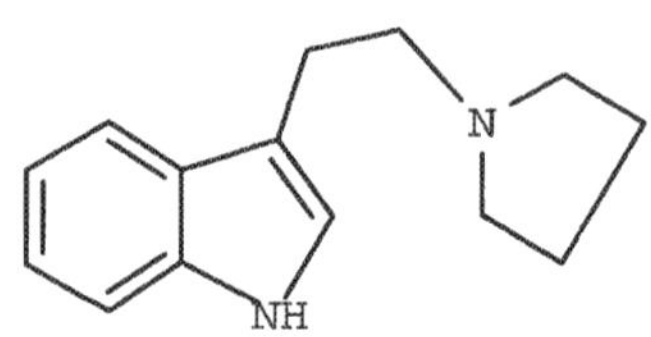

MF: C14H18N2
MW: 214.31
CA Registry Number: [14008-96-9]; hydrochloride [14009-37-1]
CA Chemical Name: Indole, 3-[2-(1-pyrrolidinyl)ethyl]-
Category: Psychotomimetic indolealkylamine
Street Names: Not reported
Abuse: Rare
Type of action: Psychotomimetic, hallucinogen
Human active dose: 50-70 mg
Duration of action: 3-5 hours
Toxic manifestations: Tachycardia, hypertension, mydriasis, anorexia, ataxia, hyperreflexia, hyperthermia, coma
Toxicity: Not reported

A seizure of this close analogue of N,N-diethyltryptamine has been reported in England [1]. The reference also describes its identification by GC-MS, IR and NMR .

References:

[1] Cowie, J.S.; Holtham, B.S.; Jones, B.S.: J. Forensic Sci., **27**, p.527, (1982)

2.4 Synthetic Cannabinoids

Due to the abundance of natural THC (**1**) on the drug market, the synthetic cannabinoids are only of very limited interest to a clandestine chemist.

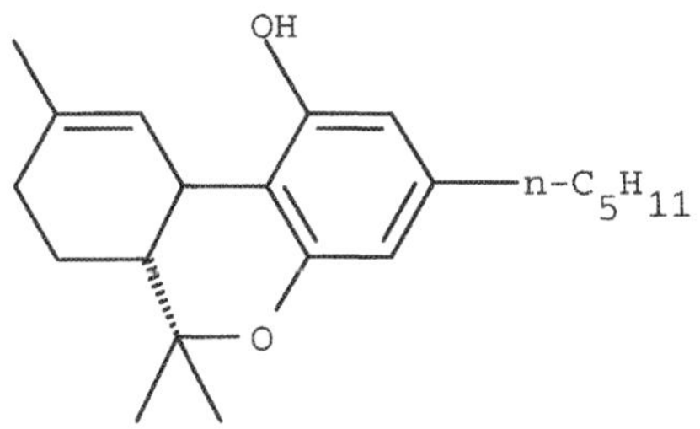

1

2-3 mg (inhaled)

Consequently, only Δ3-THC (**2**) [1] which is easy to prepare from commercially available precursors is sporadically manufactured.

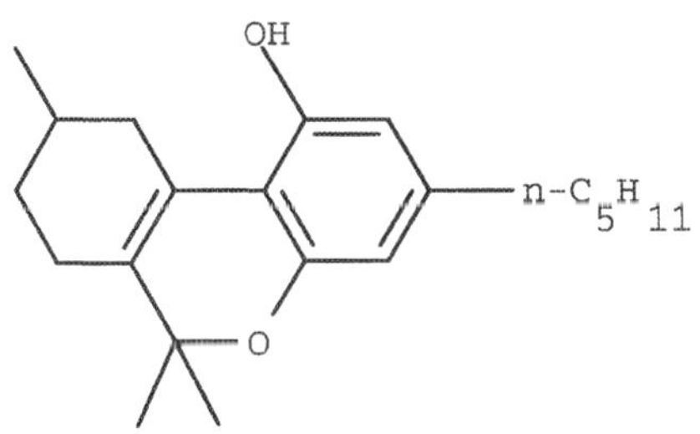

2

10-15 mg (inhaled)

However, a partial synthesis of THC isomers from cannabis resin obtained by extraction of industrial hemp or poor marihuana has been thoroughly described in the clandestine literature. This resin usually contains a high concentration of inactive cannabidiol (**3**) and THC-COOH. These compounds are converted into highly active THC isomers by cyclisation and decarboxylation, respectively. These reactions take place at elevated temperature and under acid catalysis. Usually, a mixture of various THC isomers is obtained in the proportion depending on the reaction conditions [2].

In the clandestine sources, this method is rather incorrectly referred to as isomerisation. Two models of apparatus specially designed for this procedure are found on the market. (Isomerizer I and II) [3].

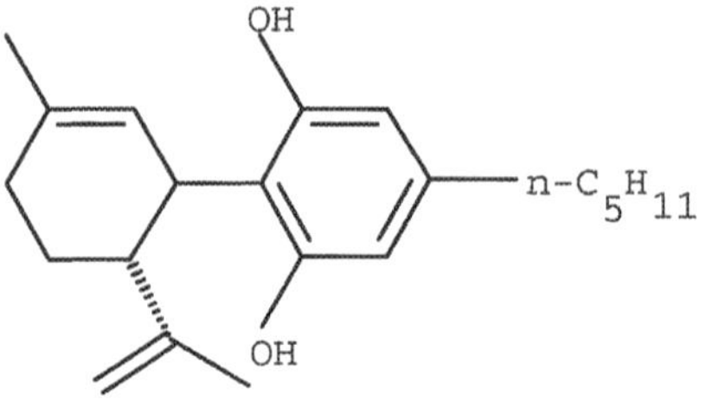

3

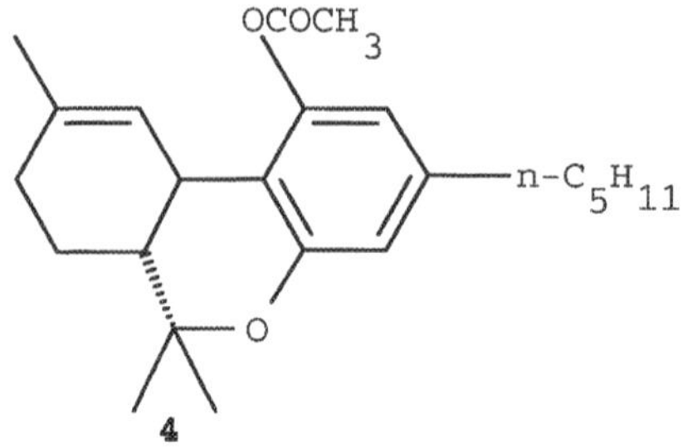

4

About 1 mg (inhaled)

Another partially synthetic method, designed to enhance the potency of natural THC, involves its acetylation because acetyl-THC (**4**) is about 3 times as active [4].

The reaction is mostly carried out in a solution using acetanhydride as an acetylating agent. According to the clandestine sources, the Isomerizer I and II apparatus, mentioned above, are also well suited for this reaction. Because it is mostly a quite complicated mixture of THC isomers, which is subjected to the treatment, the resulting product is a still more complex combination of acetylated THC isomers. Consequently, these latter compounds may be difficult to identify.

An excellent review on the structure-activity relationships of natural and synthetic cannabinoids has been published [5].

2.4.1 General References

[1] Smith, M.V.: Psychedelic Chemistry, Loompanics Unlimited, Box 1197, Port Townsend, WA 98368, p.16, (1981)
[2] ibid., pp.9-10
[3] Gold, D.: Cannabis Alchemy, Ronin Publishing, Inc., Box 1035, Berkeley, CA 94701, pp.17 and 63, (1990)
[4] ibid., p.24
[5] Duane Sofia, R.: Cannabis: Structure-Activity Relationships. In: Handbook of Psychopharmacology, Vol.**12**, Eds. Iversen, L.L.; Iversen, S.D.; Snyder, S.H.; Plenum Press, New York, pp.337-343, (Chapter 6), (1978)

2.4.2 Data Sheets: Synthetic Cannabinoids

Substance Number 2 in Chapter 2.4

IUPAC Name: 6,6,9-Trimethyl-3-pentyl-7,8,9,10-tetrahydro-6H-benzo[c]chromen-1-ol
Synonyms: 7,8,9,10-Tetrahydro-6,6,9-trimethyl-3-pentyl-6H-dibenzo[b,d]pyran-1-ol
Δ3-THC
Δ6a,10a-THC

Chemical Structure:

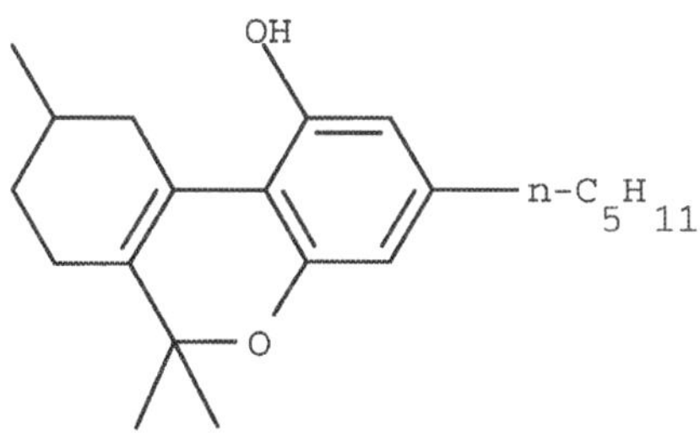

MF: C21H30O2
MW: 314.467
CA Registry Number: [7663-50-5]; [R]-[95720-01-7]; [S]-[95720-02-8]
CA Chemical Name: 6H-Dibenzo[b,d]pyran-1-ol, 7,8,9,10-tetrahydro-6,6,9-trimethyl-3-pentyl-
Category: Synthetic cannabinoid
Street Names: 3-THC
Abuse: Rare
Type of action: Psychotomimetic
Human active dose: 10-15 mg (inhaled)
Duration of action: 3-5 hours
Toxic manifestations: Tachycardie, hypotension
Toxicity: Not reported

This compound is practically the only synthetic THC analogue which appears sporadically in the clandestine market because it can be prepared easily, in a single step, from commercially available precursors [1]. This reference also describes the clandestine synthesis of much more powerful Δ3-THC analogues which are well known in the open literature [2]. Due to the need for complicated precursors, they are very unlikely to appear in illicit traffic. The identification of this substance by GC and microchemical methods has been reported [3].

References:

[1] Smith, M.V.: Psychedelic Chemistry, Loompanics Unlimited, Box 1197, Port Townsend, WA 98368, p.16, (1981)
[2] Duane Sofia, R.: Cannabis: Structure-Activity Relationships. In: Handbook of Psychopharmacology, Vol.**12**, Eds. Iversen, L.L.; Iversen, S.D.; Snyder, S.H.; Plenum Press, New York, pp.337-343, (Chapter 6), (1978)
[3] Shimamine, M.; Takahashi, K.; Ono, M.: Eisei Shikensho Hokoku, **97**, p.82, (1979), (in Japanese); CA **93**:101531

Substance Number 4 in Chapter 2.4

IUPAC Name: Acetic acid, 6,6,9-trimethyl-3-pentyl-6a,7,8,10a-tetrahydro-6H-benzo[c]chromen-1-yl ester (6aR-trans)
Synonyms: 6a,7,8,10a-Tetrahydro-6,6,9-trimethyl-3-pentyl-6H-dibenzo[b,d]pyran-1-ol acetate, (6aR-trans)
Acetyl-Δ9-THC

Chemical Structure:

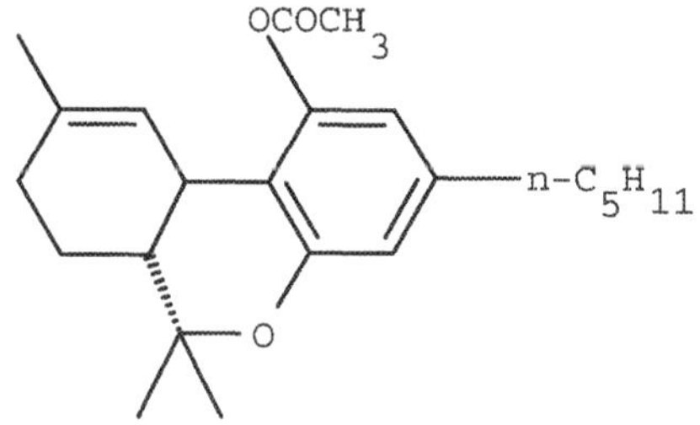

MF: C23H32O3
MW: 356.504
CA Registry Number: [23132-17-4]
CA Chemical Name: 6H-Dibenzo[b,d]pyran-1-ol, 6a,7,8,10a-tetrahydro-6,6,9-trimethyl-3-pentyl-acetate, (6aR-trans)-
Category: Synthetic cannabinoid
Street Names: Superhash
Abuse: Limited
Type of action: Psychotomimetic, hallucinogen
Human active dose: About 1 mg (inhaled)
Duration of action: 2-4 hours
Toxic manifestations: Tachycardie, hypotension
Toxicity: LD50 is 27.5 mg/kg i.v. (rat).

Prepared from natural THC, this drug is actually a partially synthetic THC analogue [1]. Consequently, a series of acetylated THC isomers are found in the acetylated product, which may be difficult to identify. For instance, these acetylated derivatives do not practically react with Fast Blue B and their TLC, GC and HPLC properties are considerably modified in comparison to the natural THC mixture. Only one case of seizure of this product was reported in the USA [2] but Internet monitoring and clandestine sources show that this treatment of poor marihuana is a common practice [1,3].

References:

[1] Gold, D.: Cannabis Alchemy, Ronin Publishing, Inc., Box 1035, Berkeley, CA 94701, p.24, (1990)
[2] Cooper D.A.: Future Synthetic Drugs of Abuse. Drug Enforcement Administration, McLean, Virginia. Text available at http://www.damicon.fi/drugs/misc/future.html, p.2
[3] http://www.hyperreal.org/drugs/marijuana/usage/isomerize

2.5 Phencyclidine and its Congeners

The drugs in this chapter are frequently classified as hallucinogens. In humans, they produce psychological disturbances that are more closely related to various psychopathological states than those shown by the classical hallucinogens. Therefore, these substances deserve their own category [1]. Ebriety, disorientation, amnesia, visual, auditory and tactile illusions and delusions are usually present. A particular feeling of allmightiness, hostility and violent behaviour also seem to be typical symptoms of the intoxication.

Due to the accompanying anesthesia, the intoxicated and disoriented persons do not perceive pain and consequently may suffer serious injuries. Physiological effects which appear at moderate or high doses include comatose state, muscular rigidity, involuntary motor activity, generalised seizures, hypertension and renal failure. A long lasting toxic psychosis and depression are frequent complications of a serious phencyclidine intoxication.

Phencyclidine analogues show a very similar pharmacological profile. Only ketamine displays a lower incidence of the disturbances described above. Clinical management of poisoning by this type of drug has been thoroughly described [1].

Originally, ketamine (**1**) [2] and phencyclidine (**2**), (Sernyl, Sernylan, PCP, « Angel dust », « Hog ») were produced by Parke-Davis, Inc., for human medical use as general anaesthetics in surgery [4].

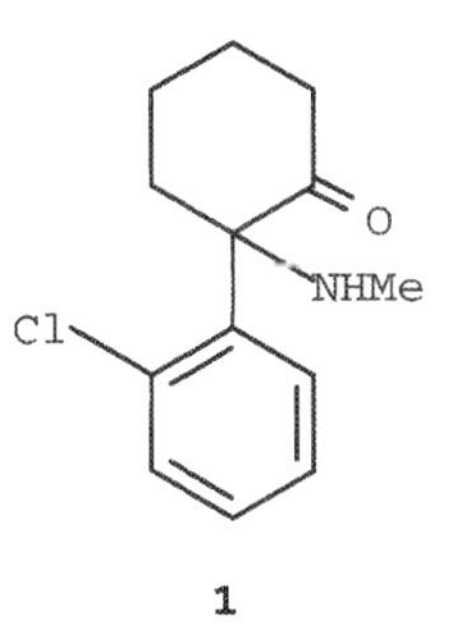

1
70-500 mg

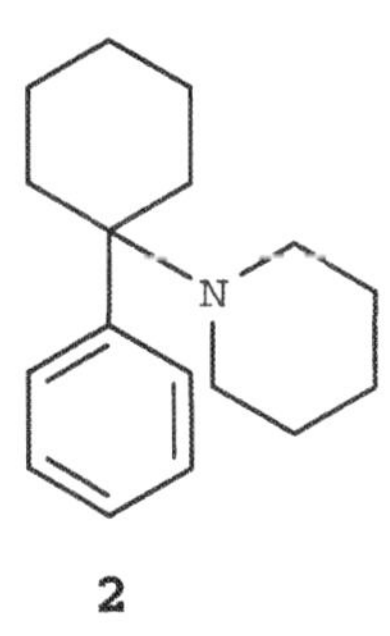

2
5-10 mg

Ketamine (Ketalar) is still utilised in medicine [5]. As it is legally produced in the USA, Hungary and Russia, this drug is readily available.

On the other hand, phencyclidine, after discovery of its serious side effects (neurotic and psychotic decompensation, important and long lasting post-operative confusion), was reserved practically only for veterinary use (Sernylan). At present, it is a controlled substance with limited medical use.

Being easy and inexpensive to manufacture, phencyclidine and its analogues very frequently replace many psychotomimetics.

According to a statistical study [6], only 3 % of the total volume of phencyclidines on the market are sold under typical phencyclidine street names which are also used for all phencyclidine analogues. The remaining 97% are marketed as THC, LSD, mescaline, MDMA etc.

Phencyclidine and its analogues [3,12] are orally active drugs but they act much more rapidly when inhaled. Consequently, smoking is the most preferred way of administration of these substances. Mostly, cannabis, parsley or oregano leaves are utilised as a support [6].

Both ketamine and phencyclidine are well known drugs of abuse and their identification, analytical and toxicological properties have been thoroughly studied.

However, the analytical and particularly toxicological properties of the phencyclidine analogues remain poorly investigated. These compounds have appeared on the clandestine market for several reasons. Similarly to phencyclidine, they can be easily and economically manufactured but their synthesis usually avoids the utilisation of highly suspect precursors. Moreover, they may be even more potent than phencyclidine itself [7,8].

Thus, the pyrrolidine PCP analogue (PHP, **3**) is as potent as phencyclidine [9] but its manufacture employs readily available pyrrolidine instead of piperidine which is a highly controlled chemical.

The ethyl analogue (PCE, **4**) [10] and the heterocyclic, short acting analogues (TCP, **5**; TCPy, **6**) belong to the most powerful compounds in this category [9].

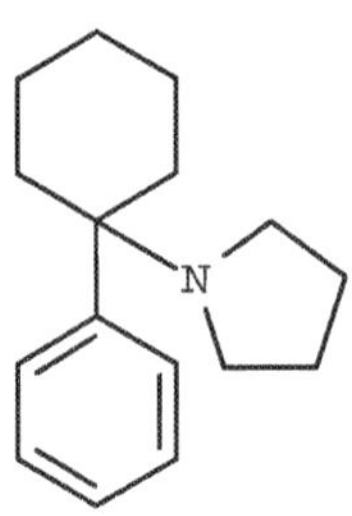

3

5-10 mg

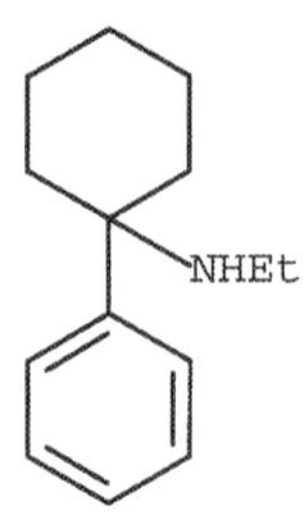

4

3-5 mg

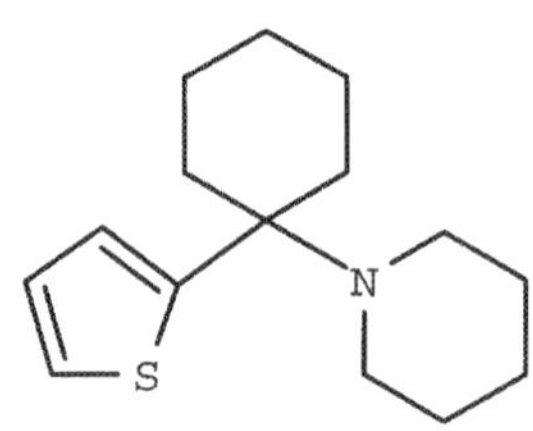

5

2-3 mg

Utilisation of readily available morpholine in the phencyclidine synthesis yields the less potent analogue (**7**). Here, the lower potency is the price paid by a clandestine manufacturer for his higher security.

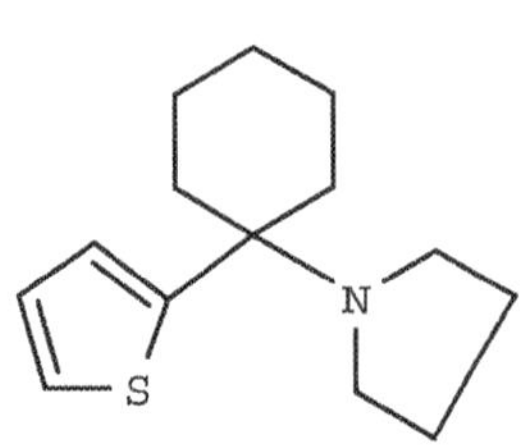

6

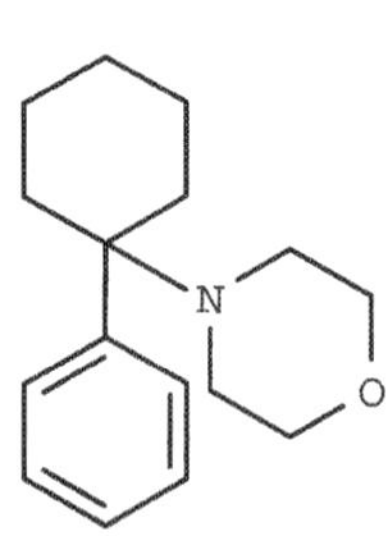

7

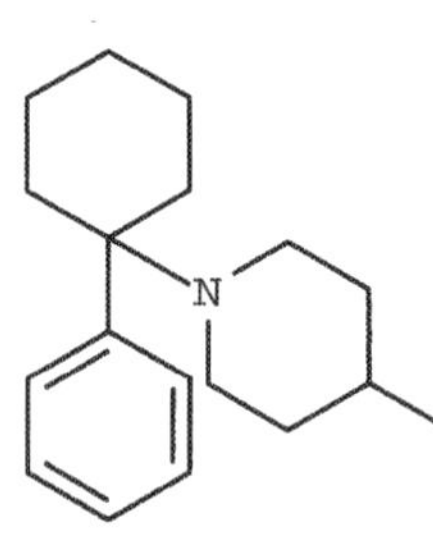

8

For the same reasons, 4-methylpiperidine was used for preparation of the analogue (**8**) which was detected in 1981 in the USA.

The 8-fold decrease in potency in comparison to phencyclidine was probably not expected by the designer of this compound [11]. Also, the 4-methylphenyl- and the benzylamine analogues of phencyclidine (**9, 10**) appear sporadically in the market [13].

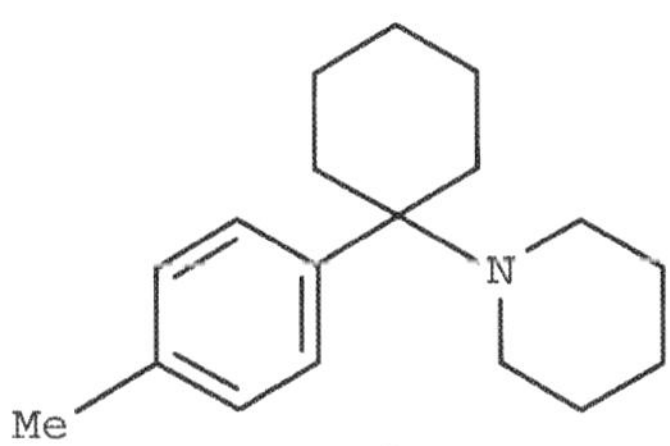

9

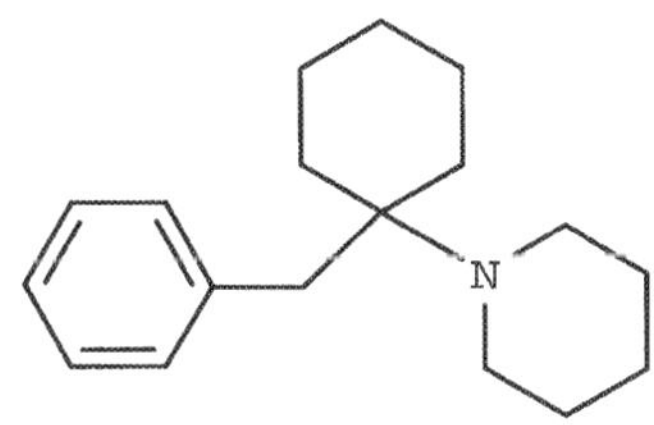

10

The synthetic routes leading to PCP and its analogues have been thoroughly reviewed [14].

2.5.1 General References

[1] Litowitz, T.L.: Phencyclidine (PCP). In: Clinical Management of Poisoning and Drug Overdose, W.B. Saunders Company, Eastbourne, UK, p. 448, (1983)
[2] Kreuscher, H. (Ed.): Ketamine. In: Anaesthesiologie und Wiederbelebung, Vol.**40**, Springer Verlag, Berlin, (1969)
[3] Maddox, V.H.; Godefroi, E.F.; Parcell, R.F.: J. Med. Chem., **8**, p.230, (1965)
[4] Marquis, K.L. et al.: Pharm. Bioch. Behav., **27**, p.385, (1987)
[5] Hansen, G. et al.: J. of Psychoactive Drugs, **20**, p.419, (1988)
[6] Ref. 1, p. 448
[7] Shulgin, A.T.; Mac Lean, D.E.: Clin. Toxicol., **9**, p.553, (1976)
[8] Helisten, C. et al.: J. Chromatogr., **117**, p.232, (1976)
[9] Ref. 1, p. 451
[10] Clark, C.C.: J. Forensic Sci., **32**, p.917, (1987)
[11] Soine, W.H. et al.: J. Analyt. Toxicol., **6**, p.41, (1982)
[12] Kalir, A. et al.: J. Med. Chem., **12**, p.473, (1977)
[13] Lodge, B.A.; Duhaime, R.; Zamecnik, J.; MacMurray, P.; Brousseau, R.: Forensic Sci. Int., **55**, p.13, (1992)
[14] Allen, A.C.; Robles, J.; Dowenski W.; Calderon, S.: Forensic Sci. Int,. **61**, p.85, (1993)

2.5.2. Data Sheets: Phencyclidine and its Congeners

Substance Number 1 in Chapter 2.5

IUPAC Name: 2-(2-Chloro-phenyl)-2-methylamino-cyclohexanone
Synonyms: 2-(2-Chlorophenyl)-2-(methylamino)cyclohexanone
Ketamine
Ketalar

Chemical Structure:

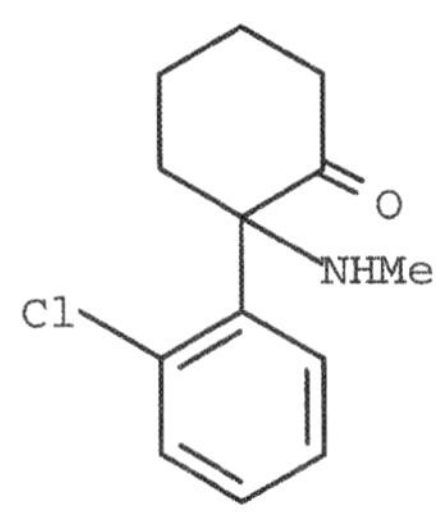

MF: C13H16NClO
MW: 237.729
CA Registry Number: Racemate [6740-88-1]
CA Chemical Name: Cyclohexanone, 2-(2-chlorophenyl)-2-(methylamino)-
Category: Phencyclidines
Street Names: Green; Purple; K; Special K; Special la coke; Super acid; Super C
Abuse: Frequent
Type of action: Anaesthetic, psychotomimetic
Human active dose: 70-500 mg (injected or inhaled)
Duration of action: About 1 hour
Toxic manifestations: Disorientation, anaesthesia, catalepsy, muscular rigidity, convulsions
Toxicity: LD50 is 224 mg/kg i.p. (mouse).

This substance is employed in human medicine as a general anaesthetic but its abuse is frequent. It appears sometimes in the « Rave » environment. In man, it produces a particular « out-of body » experience (at doses from 200 to 500 mg) [1], that has even led to the creation of a new philosophical current. The intoxication is usually short but impaired judgement and coordination may last for 18 to 24 hours [1]. A recent review on the psychic effect of ketamine has been published [2]. This drug actually does not completely deserve the name "designer drug" because there is no evidence of its clandestine manufacture. A GC based method for the identification of ketamine has been published [3].

References:

[1] Lawrence, J.: Microgram, XXIX, (8), p.202, (1996)
[2] Hansen, G.; Jensen, S.B.; Chandresh, L.; Hilden, T.: J. of Psychoactive Drugs, **20**, p.419, (1988)
[3] Drummer, O.H.; Horomidis, S.; Kourtis, S.; Syrjanen, M.L.; Tippett, P.: J. Anal. Toxicol., **18**, p.134, (1994)

Substance Number 2 in Chapter 2.5
IUPAC Name: 1-(1-Phenyl-cyclohexyl)-piperidine
Synonyms: 1-(1-Phenylcyclohexyl)piperidine
Phencyclidine

Chemical Structure:

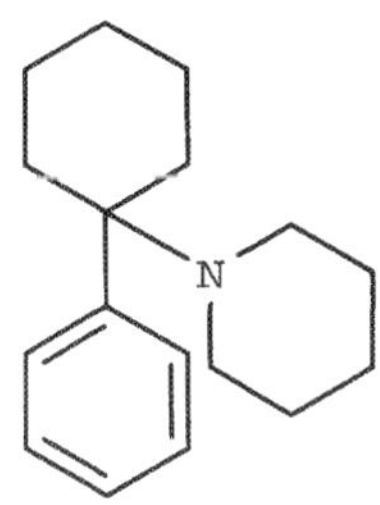

MF: $C_{17}H_{25}N$
MW: 243.391
CA Registry Number: [77-10-1]; hydrochloride [956-90-1]
CA Chemical Name: Piperidine, 1-(1-phenylcyclohexyl)-
Category: Phencyclidines
Street Names: Angel dust; Hog; PCP; PeaCe Pill; Cadillac; Crystal
Abuse: Very frequent
Type of action: Anaesthetic, psychotomimetic
Human active dose: 5-10 mg
Duration of action: 4-6 hours
Toxic manifestations: Disorientation, anaesthesia, catalepsy, muscular rigidity, convulsions
Toxicity: LD50 is 76.5 mg/kg p.o. (mouse). In humans, the plasma levels in the range of 2-2.5 µg/mL are lethal.

This drug [1] appeared for the first time in California in 1967. It used to be distributed, before discovery of its abuse potential, as a veterinary anaesthetic (Sernylan). Later the substance become controlled and its clandestine production began immediately. The identification of this drug by MS and HPLC has been described [2].

The popularity of this drug is best illustrated by the number of street names listed below. These names are also used for all phencyclidine analogues, except ketamine:

AD; Amoeba; Angel dust; Angel hair; Angel mist; Angel Poke; Angel; Animal trank; Animal tranquilizer; Aurora borealis; Blud madman; Boat; Busy bee; Butt naked; Cadillac; Cigarrode cristal; CJ; Cliffhanger; Cycline; Cyclones; Crystal T; Detroit pink; Dipper; Dirge; Do it Jack; DOA; Drink; Dummy dust; Dust joint; Dust of angels; Dusted parsley; Elephant; Flakes; Fresh; Good; Goon dust; Goon; Gorilla biscuits; Gorilla tab; HCP; Heaven and Hell; Herms; Hog; Jet fuel; K-blast; Kaps; KJ; Kay Jay; Koller joints;

Kools; Krystal joint; Krystal; KW; Leaky bolla; Leaky leak; Lovely; Mad dog; Madman; Magic dust; Magic; Mean green; Mint leaf; Mint weed; Monkey dust; New acid; New magic; O.P.P.; Ozone; P; Paz; PCP; PCPA; Peace pill; Peace; Peep; Pit; Polvo de angel; Polvo de estrellas; Rocket fuel; Rupture; Scaffle; Selma; Sherman; Soma; Stardust; Super joint; Super kools; Super weed; Surfer; T; TAC; T-buzz; Zombie dust.

References:

[1] Peterson, R.C.; Stillman, R.C. (eds.): Phencyclidine (PCP) Abuse: An Appraisal. NIDA Research Monograph, **21**, (1978)
[2] Mills III, T.; Roberson, J.C.; McCurdy, H.H.; Wall, W.H.: Instrumental Data for Drug Analysis, Elsevier Science Publishing Co., Inc., p.1780, (1992)

Substance Number 3 in Chapter 2.5

IUPAC Name: 1-(1-Phenyl-cyclohexyl)-pyrrolidine
Synonyms: 1-(1-Phenylcyclohexyl)pyrrolidine
PCPy
Rolicyclidine

Chemical Structure:

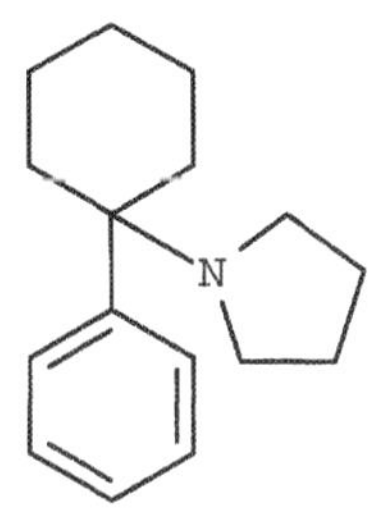

MF: C16H23N
MW: 229.364
CA Registry Number: [2201-39-0]
CA Chemical Name: Pyrrolidine, 1-(1-phenylcyclohexyl)-
Category: Phencyclidines
Street Names: PHP; Angel dust, etc.
Abuse: Frequent
Type of action: Anaesthetic, psychotomimetic
Human active dose: 5-10 mg
Duration of action: 4-6 hours
Toxic manifestations: Disorientation, anaesthesia, catalepsy, muscular rigidity, convulsions
Toxicity: Not reported

This drug [1] has practically the same pharmacological profile as phencyclidine [2]. It is produced to avoid the use of piperidine. This phencyclidine precursor is a strictly controlled chemical while pyrrolidine is not so closely watched [2]. The identification of this drug by MS and HPLC has been described [3].

References:

[1] Kalir, A.; Edery, H.; Pelah, Z.; Balderman, D.; Porath, G.: J. Med. Chem., **12**, p.473, (1977)
[2] Litovitz, T.: Phencyclidine (PCP). In: Clinical Management of Poisoning and Drug Overdose, Eds. Haddad, L.M.; Winchester, J.F.; W.B. Saunders Company, Eastbourne, UK, p.451, (1983)
[3] Mills III, T.; Roberson, J.C.; McCurdy, H.H.; Wall, W.H.: Instrumental Data for Drug Analysis, Elsevier Science Publishing Co., Inc., p.1788, (1992)

Substance Number 4 in Chapter 2.5

IUPAC Name: Ethyl-(1-phenyl-cyclohexyl)-amine
Synonyms: N-Ethyl-1-phenylcyclohexylamine
Ethylphencyclidine

Chemical Structure:

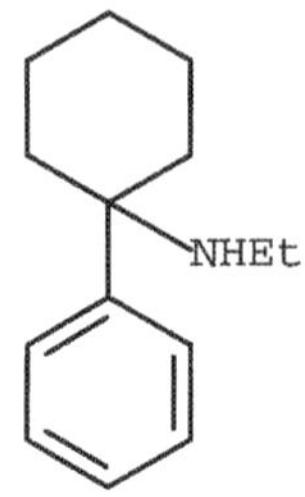

MF: C14H21N
MW: 249.413
CA Registry Number: [2201-15-2]
CA Chemical Name: Cyclohexanamine, N-ethyl-1-phenyl-
Category: Phencyclidines
Street Names: PCE; Angel dust, etc.
Abuse: Frequent
Type of action: Anaesthetic, psychotomimetic
Human active dose: 3-5 mg
Duration of action: 5-7 hours
Toxic manifestations: Disorientation, anaesthesia, catalepsy, muscular rigidity, convulsions
Toxicity: Not reported

This drug [1] has practically the same pharmacological profile as phencyclidine [2]. Again, this substance is produced to avoid the use of strictly controlled piperidine, which is a precursor necessary in phencyclidine synthesis. Substitution of more readily available ethylamine for piperidine in the synthesis yields still a more potent drug than phencyclidine itself [2]. The identification of this drug by MS and HPLC has been described [3]. Several analogues of this substance (N-Methyl-1-phenylcyclohexyl-amine, N-propyl-1-phenyl-cyclohexylamine and N-hydroxyethyl-1-phenylcyclohexylamine) have also been encountered in clandestine traffic [4].

References:

[1] Kalir, A.; Edery, H.; Pelah, Z.; Balderman, D.; Porath, G.: J. Med. Chem., **12**, p.473, (1977)

[2] Litovitz, T.: Phencyclidine (PCP). In: Clinical Management of Poisoning and Drug Overdose, Eds. Haddad, L.M.; Winchester, J.F.; W.B. Saunders Company, Eastbourne, UK, p.451, (1983)
[3] Mills III, T.; Roberson, J.C.; McCurdy, H.H.; Wall, W.H.: Instrumental Data for Drug Analysis, Elsevier Science Publishing Co., Inc., p.1782, (1992)
[4] Allen, A.C.; Robles, J.; Dowenski W.; Calderon, S.: Forensic Sci. Int., **61**, p.89, (1993)

Substance Number 5 in Chapter 2.5

IUPAC Name: 1-(1-Thiophen-2-yl-cyclohexyl)-piperidine
Synonyms: 1-[1-(2-Thienyl)cyclohexyl]piperidine
Thienylphencyclidine

Chemical Structure:

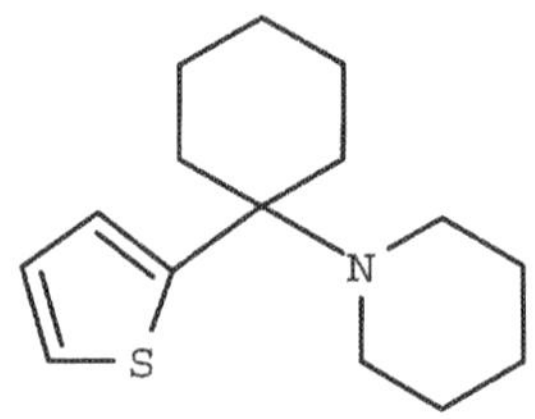

MF: $C_{15}H_{23}NS$
MW: 249.413
CA Registry Number: [21500-98-1]
CA Chemical Name: Piperidine, 1-(1-(2-thienyl)cyclohexyl)-
Category: Phencyclidines
Street Names: TCP; Angel dust etc.
Abuse: Frequent
Type of action: Anaesthetic, psychotomimetic
Human active dose: 2-3 mg
Duration of action: 3-4 hours
Toxic manifestations: Disorientation, anaesthesia, catalepsy, muscular rigidity, convulsions
Toxicity: Not reported

This drug [1] has practically the same pharmacological profile as phencyclidine but it is more powerful and shorter acting [2]. The identification of this drug by MS and HPLC has been described [3].

References:

[1] Kalir, A.; Edery, H.; Pelah, Z.; Balderman, D.; Porath, G.: J. Med. Chem., **12**, p.473 (1977)
[2] Litovitz, T.: Phencyclidine (PCP). In: Clinical Management of Poisoning and Drug Overdose, Eds. Haddad, L.M.; Winchester, J.F.; W.B. Saunders Company, Eastbourne, UK, p.451, (1983)
[3] Mills III, T.; Roberson, J.C.; McCurdy, H.H.; Wall, W.H.: Instrumental Data for Drug Analysis, Elsevier Science Publishing Co., Inc., p.2222, (1992)

Substance Number 6 in Chapter 2.5

IUPAC Name: 1-(1-Thiophen-2-yl-cyclohexyl)-pyrrolidine
Synonyms: 1-[1-(2-Thienyl)cyclohexyl]pyrrolidine
TCPy

Chemical Structure:

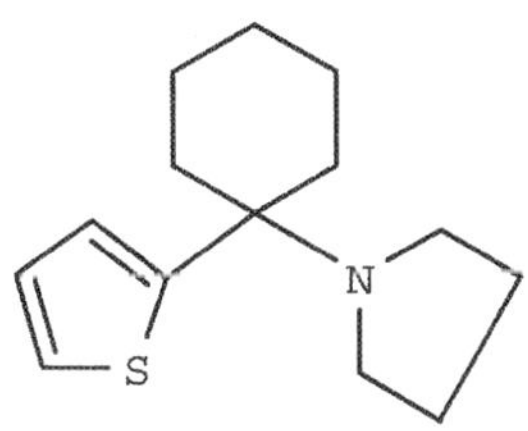

MF: C14H21NS
MW: 235.386
CA Registry Number: [22912-13-6]
CA Chemical Name: Pyrrolidine, 4-(1-(2-thienyl)cyclohexyl)-
Category: Phencyclidines
Street Names: Angel dust etc.
Abuse: Limited
Type of action: Anaesthetic, psychotomimetic
Human active dose: 5-10 mg
Duration of action: 3-4 hours
Toxic manifestations: Disorientation, anaesthesia, catalepsy, muscular rigidity, convulsions
Toxicity: Not reported

This drug [1] has practically the same pharmacological profile as phencyclidine. Again, this substance is produced to avoid the use of strictly controlled piperidine. Substitution of more readily available pyrrolidine for piperidine in the synthesis yields a compound roughly as potent as phencyclidine itself [1]. The identification of this drug by MS and HPLC has been described [2].

References:

[1] Kalir, A.; Edery, H.; Pelah, Z.; Balderman, D.; Porath, G.: J. Med. Chem., **12**, p.473, (1977)
[2] Mills III, T.; Roberson, J.C.; McCurdy, H.H.; Wall, W.H.: Instrumental Data for Drug Analysis, Elsevier Science Publishing Co., Inc., p.2224, (1992)

Substance Number 7 in Chapter 2.5

IUPAC Name: 4-(1-Phenyl-cyclohexyl)-morpholine
Synonyms: 4-(1-Phenylcyclohexyl)morpholine

Chemical Structure:

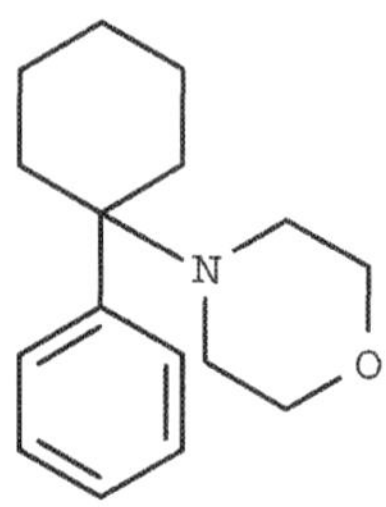

MF: C16H23NO
MW: 245.364
CA Registry Number: [2201-40-3]
CA Chemical Name: Morpholine, 4-(1-phenylcyclohexyl)-
Category: Phencyclidines
Street Names: Angel dust etc.
Abuse: Frequent
Type of action: Anaesthetic, psychotomimetic
Human active dose: 20-30 mg
Duration of action: Not reported
Toxic manifestations: Disorientation, anaesthesia, catalepsy, muscular rigidity, convulsions
Toxicity: Not reported

This drug [1] has practically the same pharmacological profile as phencyclidine. Again, this substance is produced to avoid the use of strictly controlled piperidine, which is a precursor necessary in phencyclidine synthesis. However, the substitution of more readily available morpholine for piperidine in the synthesis yields a slightly less potent drug than phencyclidine itself [1]. The identification of this drug by MS and HPLC has been described [2].

References:

[1] Maddox, V.H.; Godefroi, E.F.; Parcell, R.F.: J. Med. Chem., **8**, p.230, (1965)
[2] Mills III, T.; Roberson, J.C.; McCurdy, H.H.; Wall, W.H.: Instrumental Data for Drug Analysis, Elsevier Science Publishing Co., Inc., p.1786, (1992)

Substance Number 8 in Chapter 2.5

IUPAC Name: 4-Methyl-1-(1-phenyl-cyclohexyl)-piperidine
Synonyms: 4-Methyl-1-(1-phenylcyclohexyl)piperidine

Chemical Structure:

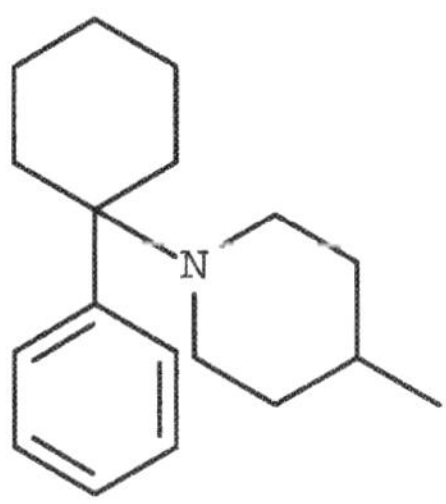

MF: C18H27N
MW: 257.418
CA Registry Number: [2201-42-5]
CA Chemical Name: Piperidine, 4-methyl-1-(1-phenylcyclohexyl)-
Category: Phencyclidines
Street Names: Angel dust etc.
Abuse: Isolated
Type of action: Anaesthetic, psychotomimetic
Human active dose: About 40 mg
Duration of action: Not reported
Toxic manifestations: Disorientation, anaesthesia, catalepsy, muscular rigidity, convulsions
Toxicity: 301.1 μmoles/kg i.p. (mice); (about 77.5 mg/kg).

This drug has practically the same pharmacological profile as phencyclidine. Again, this substance was created to avoid the use of strictly controlled piperidine.

Surprisingly, the substitution of uncontrolled 4-methylpiperidine for piperidine in the synthesis yields a much less potent, but almost as toxic drug as phencyclidine itself [1]. This reference also describes the identification of this substance by GC-MS, NMR and IR spectra.

References:

[1] Soine, W.H. et al.: J. Analyt. Toxicol., **6**, p.41, (1982)

Substance Number 9 in Chapter 2.5

IUPAC Name:1-(1-p-Tolyl-cyclohexyl)-piperidine
Synonyms: 1-(1-(4 Methylphenyl)cyclohexyl)piperidine
4-Methylphencyclidine

Chemical Structure:

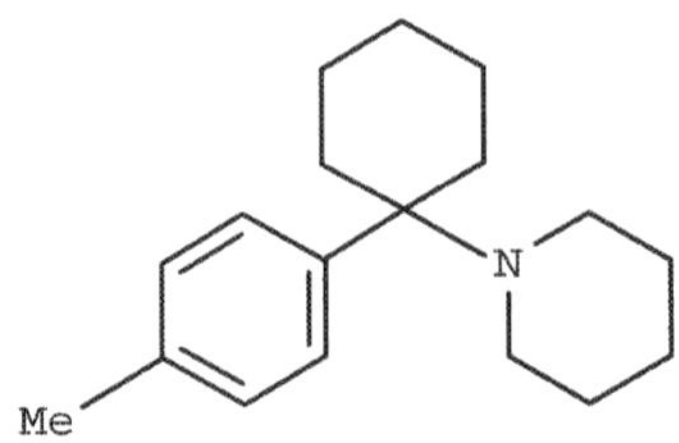

MF: C18H27N
MW: 257.418
CA Registry Number: [3883-17-8]
CA Chemical Name: Piperidine, 4-(1-(4-methylphenyl)cyclohexyl)-
Category: Phencyclidines
Street Names: Angel dust etc.
Abuse: Rare
Type of action: Anaesthetic, psychotomimetic
Human active dose: Not reported
Duration of action: Not reported
Toxic manifestations: Disorientation, anaesthesia, catalepsy, muscular rigidity, convulsions.
Toxicity: Not reported

A seizure of this phencyclidine analogue has been reported [1]. In this reference, the complete identification of this drug by GC-MS, UV, IR, NMR and C13-NMR is thoroughly described.

References:

[1] Lodge, B.A.; Duhaime, R.; Zamecnik, J.; MacMurray, P.; Brousseau, R.: Forensic Sci. Int., **55**, p.13, (1992)

Substance Number 10 in Chapter 2.5

IUPAC Name: 1-(1-Benzyl-cyclohexyl)-piperidine
Synonyms: 1-(1-Benzylcyclohexyl)piperidine

Chemical Structure:

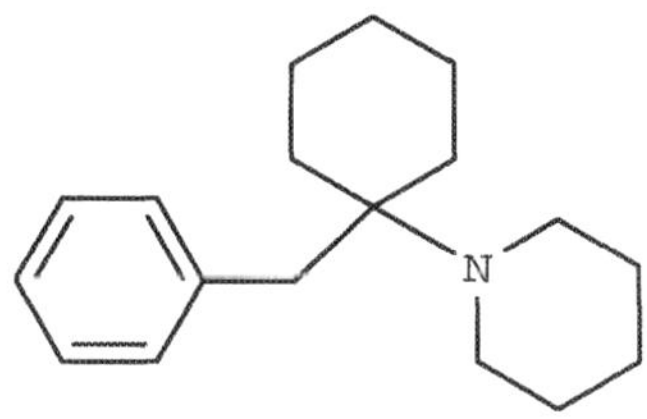

MF: C18H27N
MW: 257.418
CA Registry Number: [22912-07-8]
CA Chemical Name: Piperidine, 4-(1-benzylcyclohexyl)-
Category: Phencyclidines
Street Names: Angel dust etc.
Abuse: Rare
Type of action: Not reported
Human active dose: Not reported
Duration of action: Not reported
Toxic manifestations: Disorientation, anaesthesia, catalepsy, muscular rigidity, convulsions
Toxicity: Not reported

A seizure of this phencyclidine analogue has been reported [1]. In this reference, the identification of this drug by GC-MS, UV, IR, NMR and C13-NMR is thoroughly described.

References:

[1] Lodge, B.A.; Duhaime, R.; Zamecnik, J.; MacMurray, P.; Brousseau, R.: Forensic Sci. Int., **55**, p.13, (1992)

2.6 Deliriants

This particular category contains centrally active anticholinergic substances producing, in man, a delirium-like state accompanied by vivid visual and auditory hallucinations. These psychological disturbances closely resemble the delirium observed in alcohol withdrawal syndrome [1,2].

In clinical experimentation, the action of these substances is usually reported as extremely unpleasant. This observation may lead to the conclusion that an abuse of these substances is excluded.

However, the abuse of benactyzine (**1**), which belongs to this class of drugs, is known. This substance is a minor tranquilliser at an average daily dose of 1x3 mg but it produces a delirium-like syndrome at about 50 mg [3]. Of course, this substance is not a designer drug because there is no evidence of its illegal manufacture.

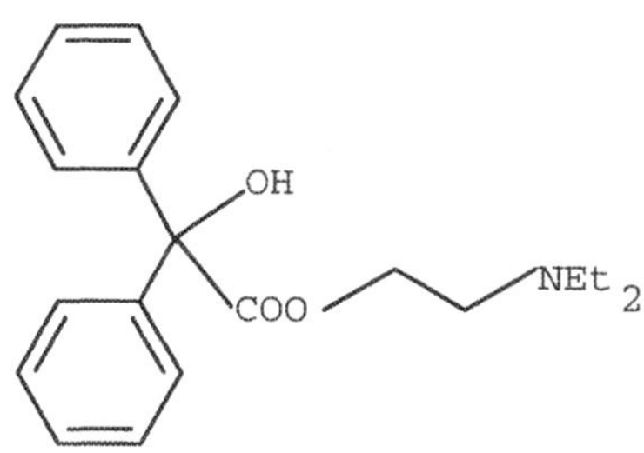

1

30-60 mg

However, there is such evidence for several other deliriants (JB336, **2** and JB-840, **3**).

In clandestine sources, the action of these drugs has been described as extraordinarily intense but too strong for current street use. A simple manufacture of these substances from uncontrolled precursors has also been correctly described [4].

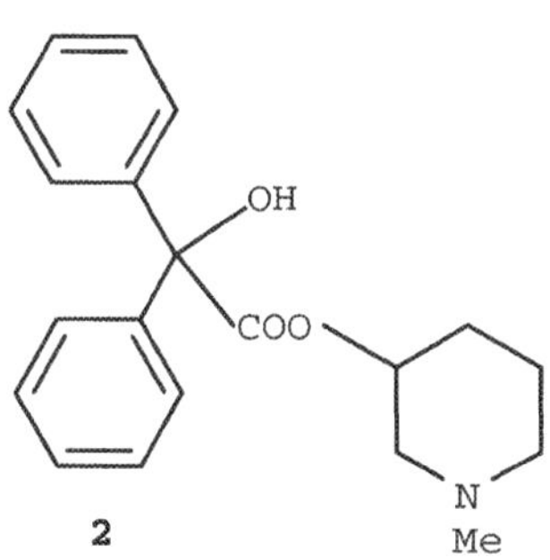

2

5-10 mg

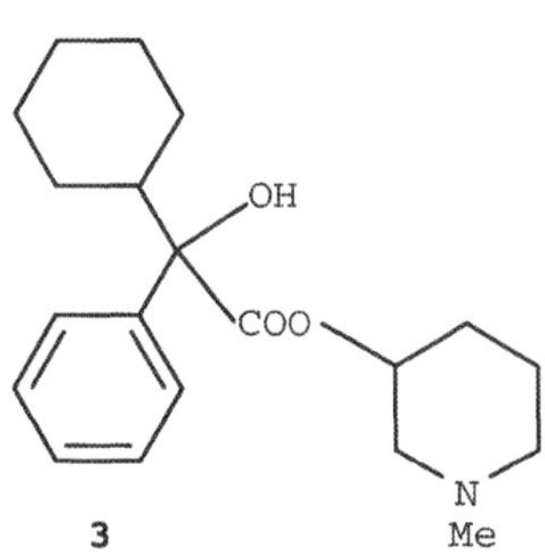

3

5-10 mg

A drug of this type was probably added to free soft drinks which were distributed by unidentified youngsters in southern France in 1996 [5].

Surprisingly, clandestine manufacturers also show an intense interest in the properties and the synthesis of the drug BZ (**4**) and its analogues [6], which also belong to the class of drugs described in this chapter.

This substance, apart from its use as a research tool in experimental pharmacology, is utilised as a chemical weapon, called the « incapacitating agent ».

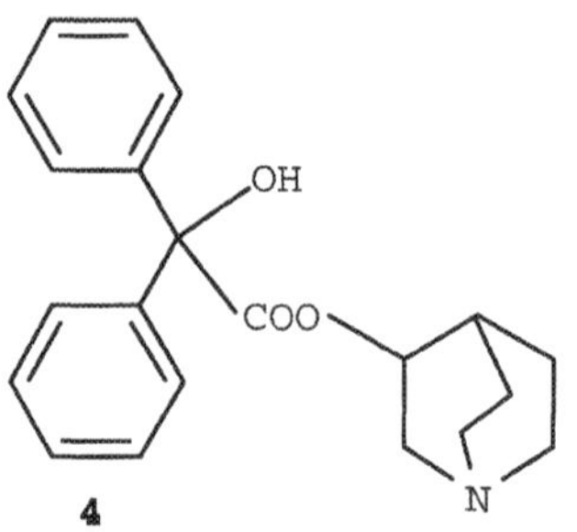

4
<1 mg

2.6.1 General References

[1] Biel, J.H. et al.: J. Am. Chem. Soc., **77**, p.2250, (1955)
[2] Abood, L.I.; Ostfeld, A.; Biel, J.H.: Arch. Int. Pharmacodyn., **120**, p.186, (1959)
[3] Grof, S. et al.: Cs. Psychiat., **6**, p.369, (1959)
[4] Smith, M.V.:Psychedelic Chemistry, Loompanics Unlimited, Box 1197, Port Townsend, WA 98368, p.167, (1981)
[5] Ugnivenko, A., M.D.: Eaux-Vives Medical Center (CMEV), Geneva, personal communication, (1996)
[6] http://www.hyperreal.org/drugs/psychedelics/misc/bz

2.6.2 Data Sheets: Deliriants

Substance Number 2 in Chapter 2.6

IUPAC Name: Hydroxy-diphenyl-acetic acid 1-methyl-piperidin-3-yl ester
Synonyms: 1-Methyl-3-piperidyl benzilate

Chemical Structure:

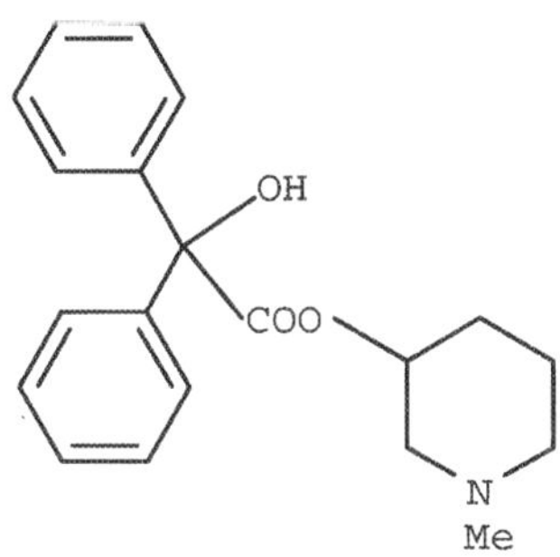

MF: C20H23NO3
MW: 325.406
CA Registry Number: [3321-80-0]
CA Chemical Name: Benzeneacetic acid, α-hydroxy-α-phenyl-, 1-methyl-3-piperidyl ester
Category: Deliriant
Street Names: JB-336
Abuse: Rare
Type of action: Deliriant
Human active dose: 5-10 mg
Duration of action: 10-12 hours
Toxic manifestations: Delirium, visual and auditory hallucinations, agitation, amnesie, catatonia, tachycardia, mydriasis
Toxicity: LD50 is 40 mg/kg, i.v. (mice) [1].

In man, this drug produces visual and auditory hallucinations and a psychological state closely resembling delirium [1]. In a clandestine manual the effect of this compound is described as very interesting but too strong for street use [2]. The compound can be easily prepared, in a single operation, from commercially available precursors [3]. The identification of the N-ethyl analogue of the title substance by GC and HPLC has been described [4].

References:

[1] Abood, L.I.; Ostfeld, A.; Biel, J.H.: Arch. Int. Pharmacodyn., **120**, p.186, (1959)

[2] Smith, M.V.: Psychedelic Chemistry, Loompanics Unlimited, Box 1197, Port Townsend, WA 98368, p.167, (1981)
[3] Biel, J.H. et al.: J. Am. Chem. Soc., **77**, p.2250, (1955)
[4] Mills III, T.; Roberson, J.C.; McCurdy, H.H.; Wall, W.H.: Instrumental Data for Drug Analysis, Elsevier Science Publishing Co., Inc., p.896, (1992)

Substance Number 3 in Chapter 2.6

IUPAC Name: Cyclohexyl-hydroxy-phenyl-acetic acid 1-methyl-piperidin-3-yl ester
Synonyms: 1-Methyl-3-piperidyl cyclohexyl phenyl glycolate

Chemical Structure:

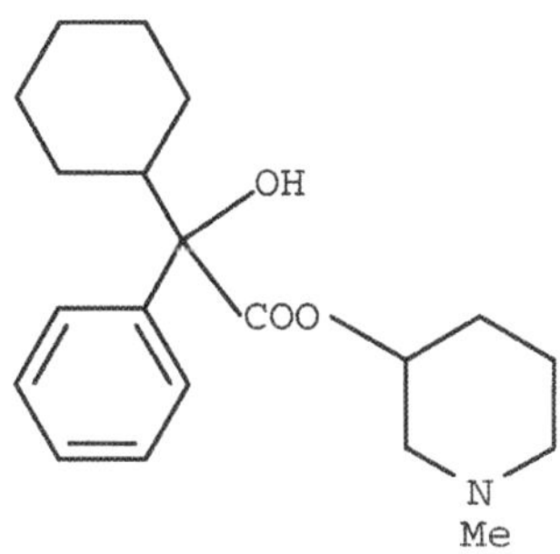

MF: C21H30NO3
MW: 347.472
CA Registry Number: [4354-45-4]
CA Chemical Name: Benzeneacetic acid, α-cyclohexyl-α-hydroxy-, 1-methyl-3-piperidinyl ester
Category: Deliriant
Street Names: JB-840
Abuse: Rare
Type of action: Deliriant
Human active dose: 5-10 mg
Duration of action: 10-14 hours
Toxic manifestations: Delirium, visual and auditory hallucinations, agitation, amnesie, catatonia, tachycardia, mydriasis
Toxicity: LD50 is 32 mg/kg, i.v. (mice) [1].

In man this drug produces visual and auditory hallucinations and a psychological state closely resembling delirium [1]. In a clandestine manual the effect of this compound is described as very interesting but too strong for street use [2]. The compound can be easily prepared, in a single operation, from commercially available precursors [3].

References:

[1] Abood, L.I.; Ostfeld, A.; Biel, J.H.: Arch. Int. Pharmacodyn., **120**, p.186, (1959)
[2] Smith, M.V.: Psychedelic Chemistry, Loompanics Unlimited, Box 1197, Port Townsend, WA 98368, p.167, (1981)
[3] Biel, J.H. et al.: J. Am. Chem. Soc., **77**, p.2250, (1955)

Substance Number 4 in Chapter 2.6

IUPAC Name: Hydroxy-diphenyl-acetic acid 1-azabicyclo[2.2.2]oct-3-yl ester
Synonyms: 3-Quinuclidinyl benzilate

Chemical Structure:

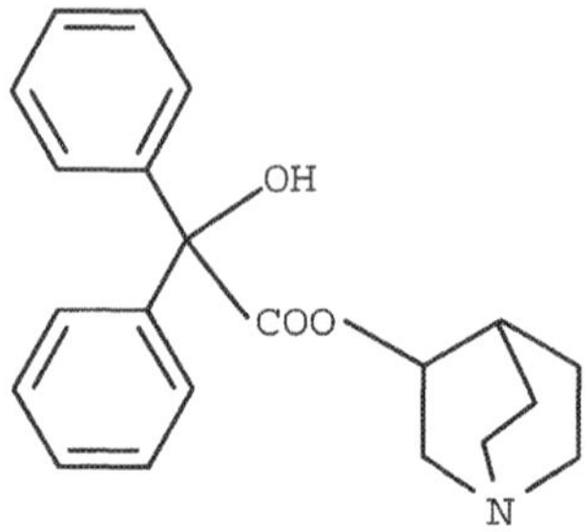

MF: C21H23NO3
MW: 337.417
CA Registry Number: [6581-06-2]; racemate [4478-53-9]; [R]-[62869-69-6]; [S]-62869-68-5]
CA Chemical Name: Benzeneacetic acid, α-hydroxy-α-phenyl-, 1-azabicyclo[2.2.2]oct-3-yl ester
Category: Deliriant
Street Names: BZ; TWA; DMZ
Abuse: Rare
Type of action: Deliriant
Human active dose: Less than 1 mg
Duration of action: Long
Toxic manifestations: Delirium, visual and auditory hallucinations, agitation, amnesia, catatonia, cardiovascular disturbances
Toxicity: LD50 is 15 mg/kg, i.v. (dog) [1].

In man this drug produces a long lasting psychological alteration [1]. For this reason, the substance was used as a chemical warfare agent (« incapacitating agent ») [3].

The compound can be easily prepared, in a single operation, from commercially available precursors. The drug appeared briefly in illicit trade in the 1960s [2], but never gained any popularity. Hence, the renewed interest in this substance shown by clandestine operators is rather surprising. Also, it remains unclear if the compound is intended to be used as an abusable drug or a weapon. The detection of the substance and of its metabolites by GC-MS has been described [3].

References:

[1] Schallek, W.; Smith, T.H.F.: J. Pharmacol. Exp. Ther., **104**, p.291, (1952)
[2] http://www.hyperreal.org/drugs/psychedelics/misc/bz
[3] Byrd, G.D.; Paule, R.C.; Sander, L.C.; Sniegoski, L.T.; White, V.E.: J. Anal. Toxicol., **16**, p.182, (1992)

2.7 CNS Stimulants

The drugs contained in this chapter exhibit rather pure CNS stimulant action in humans. This response includes mood elevation, euphoria and increase in psychomotor activity. Restlessness, irritability, confusion, hyperreflexia, hypertension and various cardiovascular disturbances may appear particularly at higher doses. Clinical toxicology and management of poisoning by this type of drug have been described [1].

Chemically, these compounds contain almost invariably a phenethylamine skeleton which may, in some cases, be part of a heterocyclic nucleus.

Ephedrine (**1**) and norephedrine (**2**) are the most frequently used precursors in the synthesis of these compounds [2]. They possess only weak CNS stimulant properties but their chemical structure can be easily modified to give a variety of more potent drugs. (Scheme 1).

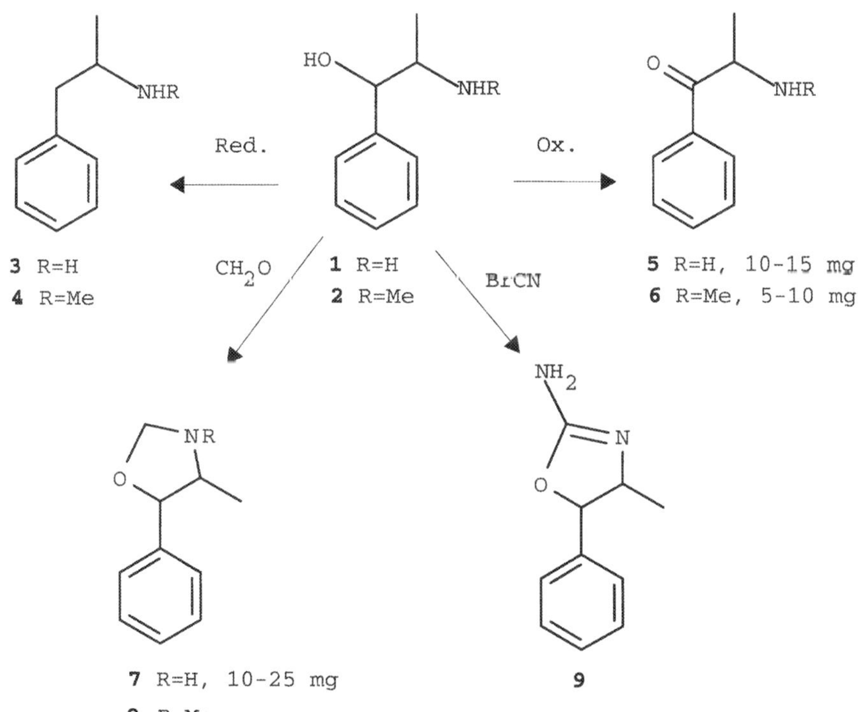

Scheme I

Thus, a hydroxyl group in ephedrine or norephedrine can be reductively eliminated. This reaction yields methamphetamine or amphetamine, respectively, which are world-wide abused drugs. An excellent review on the reduction methods employed in clandestine laboratories has been published [3].

Likewise, on oxidation, the elimination of the hydroxyl group in norephedrine and ephedrine yields cathinone (**5**) and ephedrone (**6**) (« Jeff», « CAT », S-(-)methcathinone, **10**), respectively. These two drugs, which are almost as potent as the corresponding amphetamines, have been abused in Russia since the 1980s [4]. Recently, these substances have been detected in the USA and Canada [5].

Interestingly, oxidation of pseudoephedrine gives only weakly active R-(+)methcathionine (**11**).

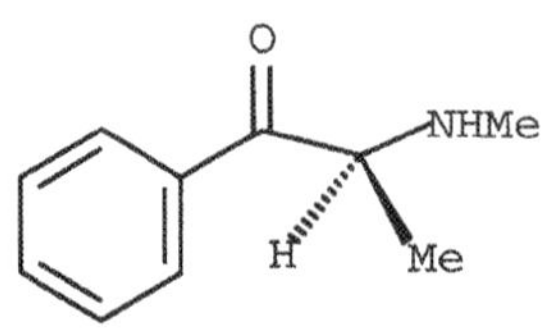

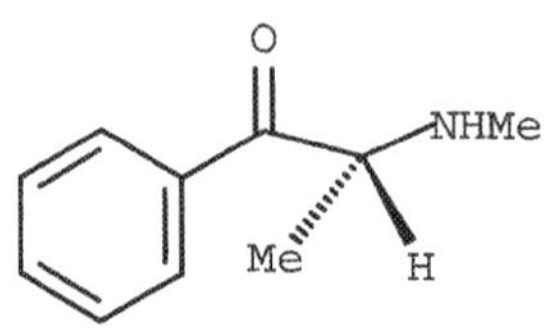

10 5-10 mg

11

Two clandestine analogues of racemic amphetamine (**12** and **13**), which can be easily prepared from phenylacetone have appeared in the illicit traffic [5]. Two other clandestine amphetamine analogues (**14** and **15**), which can be prepared from uncontrolled precursors, have also been reported [12,13].

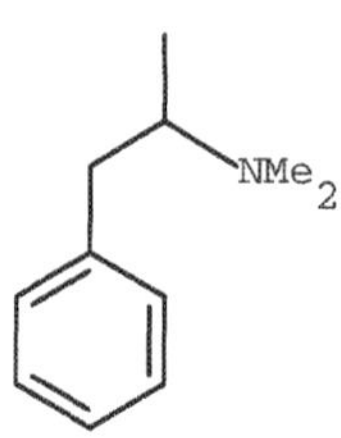

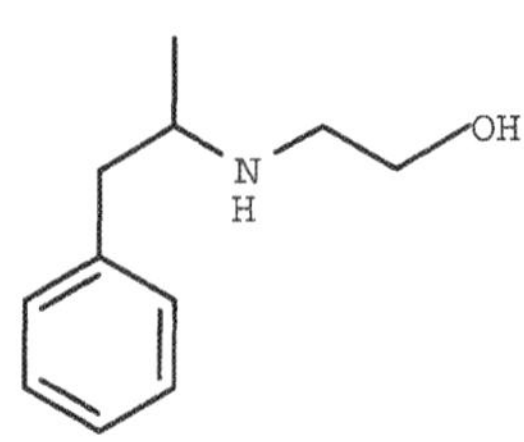

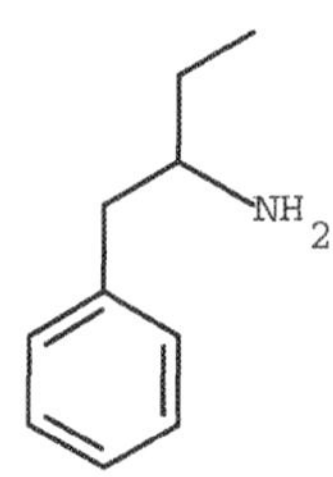

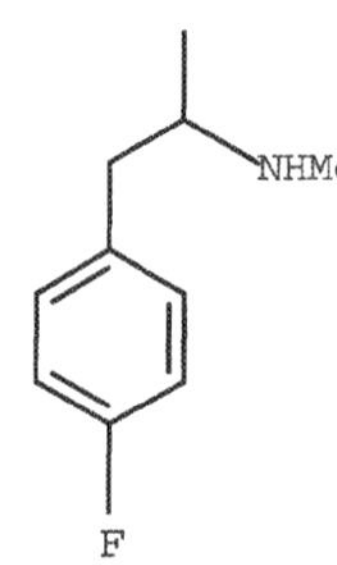

12 25-40 mg

13

14 20-40 mg

15

Cyclisation reactions may also be used to modify the chemical structure of ephedrine and its diastereomers to enhance their CNS stimulant activity. Thus, the cyclisation reaction of norephedrine with cyanogen bromide yields 4-methylaminorex (« Ice », « 4-Euph », « Euphoria », **9**) [6]. Its 4-desmethyl derivative (aminorex, **16**) is also a known designer drug [7].

In this reaction, pseudoephedrine gives an isomer of 4-aminorex which is also a potent CNS stimulant. Being strong euphoriants, these latter drugs frequently appear at the « Rave » scene.

Another extremely easy cyclisation of norephedrine and ephedrine gives oxazolidine derivatives possessing CNS stimulant properties (**7** and **8**) [8].

Methylphenidate and pipradol represent another group of CNS stimulants. They are frequently abused drugs but their clandestine manufacture have not yet been reported.

However, a pipradol analogue (**17**) and its clandestine manufacture has been described [9]. This drug is particularly easy to produce from readily available proline.

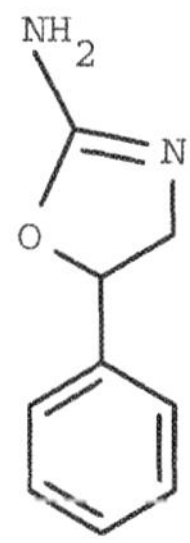

16

15-25 mg

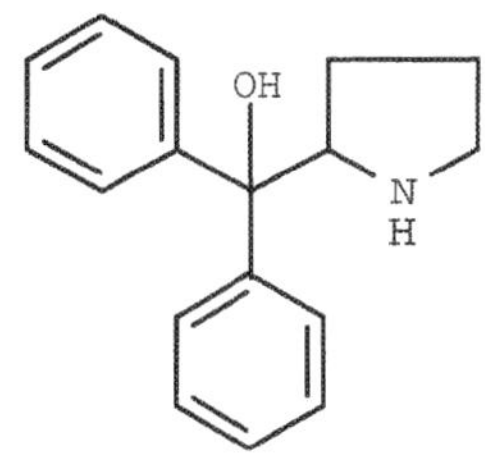

17

2-5 mg

Large quantities of α-phenethylamines (**18** and **19**) have repeatedly been seized in the clandestine drug market in Holland [10,11].

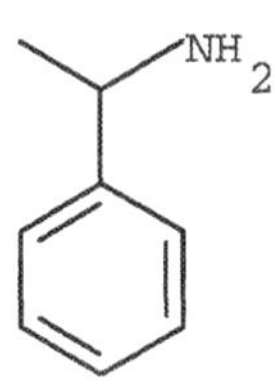

18

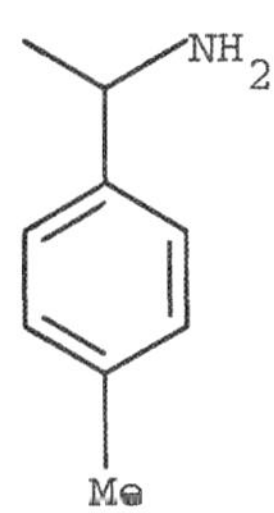

19

The aim of their clandestine production and distribution remains unclear. The substances show a very weak CNS stimulating action (about 5 to 7 fold weaker than amphetamine) but their other pharmacological properties have not been investigated. It is possible that the substances have some unknown abuse properties.

Oxazolidinone derivatives, possessing CNS stimulating properties, such as pemoline (**20**) are well known drugs of abuse but none of their analogues has appeared on the clandestine market.

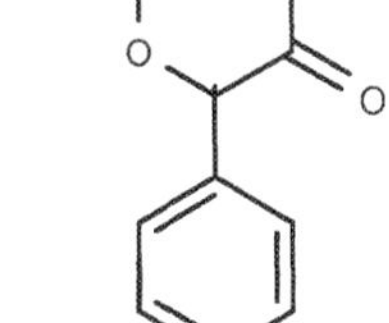

20

2.7.1 General References

[1] Litovitz, T.: Amphetamines. In: Clinical Management of Poisoning and Drug Overdose, Eds. Haddad, L.M.; Winchester, J.F.; W.B. Saunders Company, Eastbourne, UK, p.469, (1983)
[2] Report of International Narcotics Control Board, paragraph 382, (1995)
[4] Zhingel, K.Y.; Dovensky, W.; Crossman, A.; Allen, A.: J. Forensic Sci., **36**, p.915, (1991)
[5] Clark, R.C. et al.: J. Chromatogr. Sci., **32**, p.552, (1994)
[6] Klein, R.F.X. et al.: J. Forensic Sci., **34**, p.962, (1989)
[7] Brewster, M.E.; Davis, F.T.: J. Forensic Sci., **36**, p.587, (1991)
[8] Kalm, J.M.: J. Org. Chem., **25**, p.1929, (1960)
[9] Winthrop. S.O.; Humber, L.G.: J. Org. Chem., **26**, p.2834, (1961)
[10] Meyer, E. et al.: Forensic Sci. Int., **76**, p.159, (1995)
[11] King. L.A.; Poortman-van der Meer, A.J.; Huizer, H.: Forensic Sci. Int., **77**, p.141, (1996)
[12] Noggle, F.T.; Clark, C.R.; Pitts-Monk, P.; DeRuiter, J.: Microgram, XXIV, p.197, (1991)
[13] Alexander, G.R.: Microgram, XXVII, (8), p.268, (1994)

2.7.2 Data Sheets: CNS Stimulants

Substance Number 5 in Chapter 2.7

IUPAC Name: 2-Amino-1-phenyl-propan-1-one
Synonyms: 2-Amino-1-phenyl-1-propanone
Norephedrone
Cathinone

Chemical Structure:

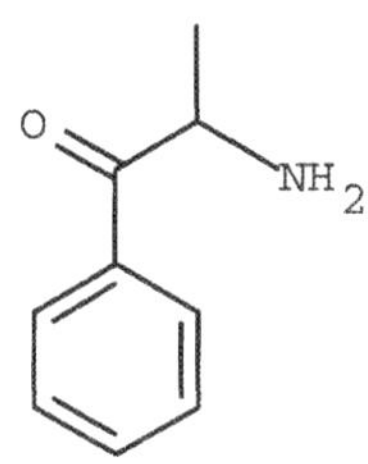

MF: C9H11NO
MW: 149.192
CA Registry Number: [5265-18-9]; racemate [75925-46-1]; [R]-[80096-54-4]; [S]-[71031-15-7]
CA Chemical Name: 1-Propanone, 2-amino-1-phenyl-
Category: CNS stimulant
Street Names: CAT
Abuse: Limited
Type of action: Stimulant
Human active dose: 10-15 mg
Duration of action: 3-4 hours
Toxic manifestations: Hyperactivity, confusion, hypertension, cardiovascular disturbances, hyperthermia, convulsions
Toxicity: Not reported

This drug of natural origin (« khat ») has been abused for a long time in the Middle East [1] but the substance of clandestine synthetic origin is likely to appear in illicit trade [2]. In man, it produces amphetamine-like CNS stimulant activity [1]. The HPLC chiral separation of this substance has recently been described [3].

References:

[1] Kalix, P.: J. of Psychoactive Drugs, **26**, p.69, (1994)
[2] Dal Cason, T.A.: Microgram, XXV, p.313, (1992)
[3] Aboul-Enein, H.Y.; Serignese, V.: Biomed. Chromatogr., **11**, p.47, (1997)

Substance Number 10 in Chapter 2.7
Unspecified, 6; S-(-)methcathinone, 10

IUPAC Name: 2-Methylamino-1-phenyl-1-propanone
Synonyms: 2-(Methylamino)-1-phenyl-1-propanone
Ephedrone
S-(-)methcathinone

Chemical Structure:

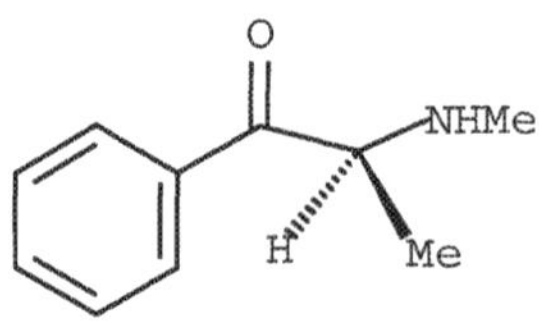

MF: C10H13NO
MW: 163.219
CA Registry Number: [5650-44-2]; racemate [28521-94-0]; [S]-[112117-24-5]
CA Chemical Name: 1-Propanone, 2-(methylamino)-1-phenyl-
Category: CNS stimulant
Street Names: Jeff; CAT
Abuse: Frequent
Type of action: Stimulant
Human active dose: 5-10 mg
Duration of action: 3-4 hours
Toxic manifestations: Hyperactivity, confusion, hypertension, cardiovascular disturbances, hyperthermia, convulsions, toxic psychosis
Toxicity: Not reported

This powerful CNS stimulant was frequently abused in Russia in the 1980's [1]. Recently (Michigan, 1990), it also appeared in the USA [2]. An overview of its epidemic in the USA [4] and methods for its identification by GC-MS [2,3] have been published.

References:

[1] Zhingel, K.Y.; Dovensky, W.; Crossman, A.; Allen, A.: J. Forensic Sci., **36**, p.915, (1991)
[2] Dal Cason, T.A.: Microgram, XXV, p.313, (1992)
[3] Calkins, R.F.; Aktan, G.B.; Hussain, K.L.: J. of Psychoactive Drugs, **27**, p.277, (1995)
[4] DeRuiter, J.; Hayes, L.; Valaer, A.K.; Clark, C.R.; Noggle, F.T.: J. Chromatogr. Sci., **32**, p.552, (1994)

Substance Number 7 in Chapter 2.7

IUPAC Name: 4-Methyl-5-phenyl-oxazolidine
Synonyms: 4-Methyl-5-phenyloxazolidine

Chemical Structure:

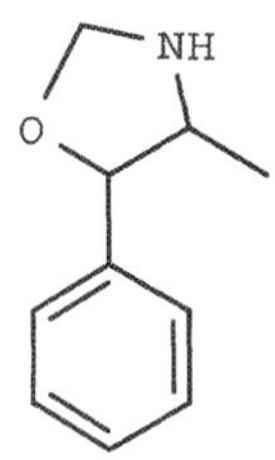

MF: C10H13NO
MW: 163.219
CA Registry Number: [42794-92-3]; cis-[70939-17-2]; trans-[70939-18-3]
CA Chemical Name: Oxazolidine, 4-methyl-5-phenyl-
Category: CNS stimulant
Street Names: Not reported
Abuse: Rare
Type of action: Stimulant
Human active dose: 10-25 mg
Duration of action: 3-4 hours
Toxic manifestations: Hyperactivity, confusion, hypertension, cardiovascular disturbances, hyperthermia, convulsions
Toxicity: Not reported

The CNS stimulant activity and synthesis of this compound was described by a sharing clandestine chemist on the Internet [1]. He also described the synthesis of the N-methyl analogue (**8**) of this substance but its active dose was omitted. The substance is known in the open literature [2].

References:

[1] Newsgroups: alt.drug.chemistry (1995)
[2] Kalm, J.M.: J. Org. Chem., **25**, p.1929, (1960)

Substance Number 9 in Chapter 2.7

IUPAC Name: 4-Methyl-5-phenyl-4,5-dihydro-oxazol-2-ylamine
Synonyms: 2-Amino-4,5-dihydro-4-methyl-5-phenyloxazole
4-Methylaminorex
4,5-Dihydro-4-methyl-5-phenyl-2-oxazolamine

Chemical Structure:

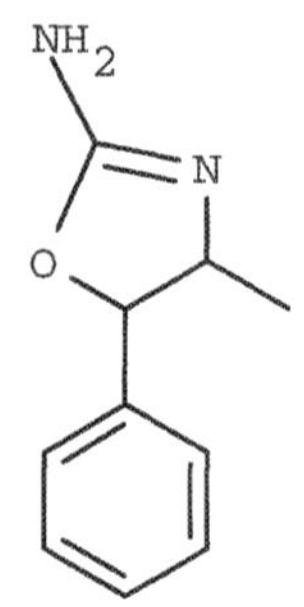

MF: C10H12N2O
MW: 176.218
CA Registry Number: [29493-77-4]; trans [2077-59-0]; [4S-trans]-[27780-30-9]; [4S-cis]-[75493-87-7]; cis, racemate [29493-77-4]; [4R-cis]-[27780-31-0]
CA Chemical Name: 2-Oxazolamine, 4,5-dihydro-4-methyl-5-phenyl-
Category: CNS stimulant
Street Names: U4-Euh; ICE; Blue Ice; Euphoria; 4-MAX
Abuse: Very frequent
Type of action: Stimulant, euphoriant, anorectic
Human active dose: 15-25 mg
Duration of action: Long
Toxic manifestations: Hyperactivity, confusion, hypertension, cardiovascular disturbances, hyperthermia, convulsions, toxic psychosis
Toxicity: Not reported

This drug is widely abused in the USA and Europe. Originally synthesized as a potent anorectic agent [1], the substance appeared on the clandestine market [2]. Being a powerful euphoriant it also appears frequently in the « Rave » environement. Apart from peroral administration, the drug is also snorted or smoked [3]. Physicochemical properties and the identification of all four possible stereoisomers of 4-methylaminorex by TLC, GC, MS, NMR and IR have been described [4]. Several studies show little difference in the pharmacology of these stereoisomers. Out of these, practically only d,l-cis 4-methylaminorex has been encountered in illicit samples [4].

References:

[1] Poos, G.; Carson, J.R.; Rosenau, J.D.; Roszkowski, A.P.; Kelley, N.M.; McGowin, J.: J. Med. Chem., **6**, p.266, (1963)
[2] Fed. Reg. **53**, No.199, p.40392; Oct. 14, (1988), Microgram, XXI, p.215, (1988)
[3] http://www.nepenthes.com/Drugs/Misc/ice.faq.html
[4] Klein, R.F.X.; Sperling, A.R.; Cooper, D.A.; Kram, T.C.: J. Forensic Sci., **34**, p.962, (1989)

Substance Number 12 in Chapter 2.7

IUPAC Name: Dimethyl-(1-methyl-2-phenyl-ethyl)-amine
Synonyms: 1-Phenyl-2-dimethylaminopropane
N,N-Dimethylamphetamine

Chemical Structure:

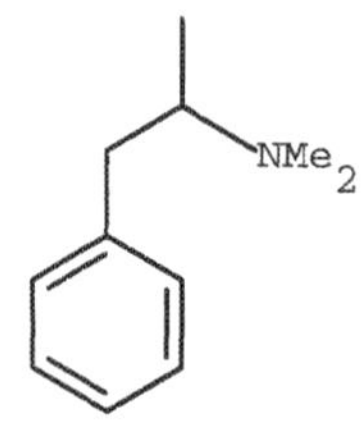

MF: C11H17N
MW: 163.30
CA Registry Number: [4075-96-1]; racemate [49681-82-5]; hydrochloride [1009-69-4]; hydrochloride racemate [33286-27-0]; [R]-[52691-87-9]; [S]-[17279-39-9]
CA Chemical Name: Benzeneethanamine, N,N,α-trimethyl-
Category: CNS stimulant
Street Names: Speed
Abuse: Frequent
Type of action: Stimulant
Human active dose: About 25-40 mg
Duration of action: Not reported
Toxic manifestations: Hyperactivity, confusion, hypertension, cardiovascular disturbances, hyperthermia, convulsions
Toxicity: Not reported

In animal studies, this drug shows about 20% amphetamine CNS stimulant activity [1]. It is illegally produced in a number of clandestine laboratories [2] and sold routinely as methamphetamine (« Speed »). The identification of this compound by GC-MS has been described [3].

References:

[1] Van der Schoot, J.B.; Ariens, E.J.; Van Rossum, J.M.; Hurkmans, J.A.: Arzneim. Forschung, **9**, p.902, (1961)
[2] Fed. Reg. **55**, No.23, p.3586, Feb. 2, (1990); Microgram, XXII, p.232, (1990)
[3] Andrews, K.M.: J. Forensic Sci., **40**, p.551, (1995)

Substance Number 13 in Chapter 2.7

IUPAC Name: 2-(1-Benzyl-ethylamino)-ethanol
Synonyms: N-(2-Hydroxyethyl)phenyl-2-propanamine
N-(2-Hydroxyethyl)amphetamine
N-(2-Hydroxyethyl)-α-methylphenethylamine

Chemical Structure:

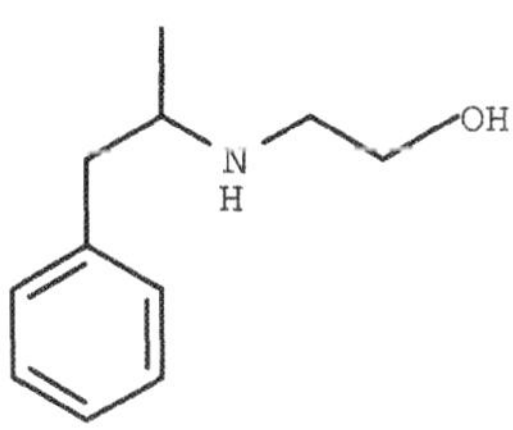

MF: C11H17NO
MW: 179.261
CA Registry Number: [63918-85-4]
CA Chemical Name: Benzeneethanamine, N-(2-hydroxyethyl)-α-methyl-
Category: CNS stimulant
Street Names: Not reported
Abuse: Rare
Type of action: Stimulant
Human active dose: Not reported
Duration of action: Not reported
Toxic manifestations: Hyperactivity, confusion, hypertension, cardiovascular disturbances, hyperthermia, convulsions
Toxicity: Not reported

This amphetamine analogue has occasionally been seized in Canada [1,2]. These references also describe the identification of this compound by GC-MS, and NMR methods.

References:

[1] Carpenter, J.; Hugel, J.; Weaver, K.: Can. Soc. Forensic Sci. J., **26**, p.143, (1993)
[2] Cyr, T.D.; Dawson, B.A.; By, A.W.; Neville, G.A.; Shurvell, H.F.: J. Forensic Sci. **41**, p.608, (1996)

Substance Number 14 in Chapter 2.7

IUPAC Name: 1-Benzyl-propylamine
Synonyms: 1-Phenyl-2-butanamine
α-Ethylphenethylamine
1-Phenyl-2-butylamine

Chemical Structure:

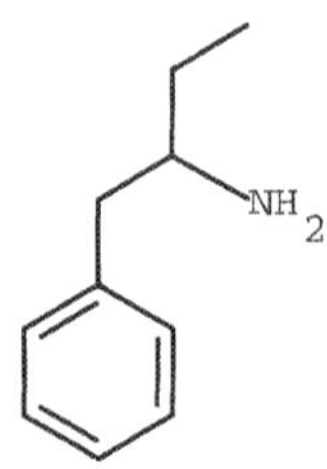

MF: C10H15N
MW: 149.235
CA Registry Number: [53309-89-0]; [S]-30543-90-9]; [R]-30543-89-6]
CA Chemical Name: Benzeneethanamine, α-ethyl-
Category: CNS stimulant
Street Names: Not reported
Abuse: Rare
Type of action: Stimulant, euphoriant
Human active dose: 20-40 mg
Duration of action: 12-16 hours
Toxic manifestations: Hyperactivity, confusion, marked and long lasting hypertension, cardiovascular disturbances, hyperthermia, convulsions
Toxicity: Not reported

This amphetamine analogue was supposed to be a designer drug in 1991 [1]. This reference also describes its identification by GC-MS. The substance possesses quite strong CNS stimulant properties [2].

References:

[1] Noggle, F.T.; Clark, C.R.; Pitts-Monk, P.; DeRuiter, J.: Microgram, XXIV, p.197, (1991)
[2] Marsh, D.F.: J. Pharmacol. Exp. Ther., **94**, p.426, (1948)

Substance Number 15 in Chapter 2.7

IUPAC Name: 2-(4-Fluoro-phenyl)-1-methyl-ethylamine
Synonyms: 4-Fluorophenyl-2-propanamine
4-Fluoromethamphetamine
N,α-Dimethyl-4-fluorophenethylamine

Chemical Structure:

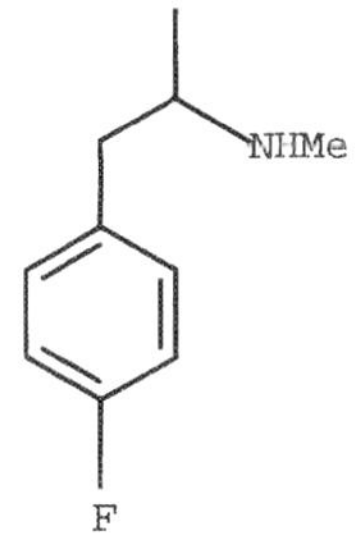

MF: C11H16FN
MW: 167.23
CA Registry Number: [351-03-1]; racemate [103596-29-8]; [R]-[103596-32-3]; [S]-[122209-64-7]; hydrochloride [R]-122290-81-7]; hydrochloride [S]-[122209-66-9]
CA Chemical Name: Benzeneethanamine, 4-fluoro-N,α-dimethyl-
Category: CNS stimulant
Street Names: Probably « Speed »
Abuse: Rare
Type of action: Stimulant
Human active dose: Not reported
Duration of action: Not reported
Toxic manifestations: Hyperactivity, confusion, hypertension, cardiovascular disturbances, hyperthermia, convulsions
Toxicity: Not reported

Attempts by clandestine operators to purchase 4-fluorophenylacetone and manufacture this drug have been reported [1]. This reference also describes the identification of this substance, including its positional isomers (2-F and 3-F) and the corresponding amphetamines by GC-MS, IR and NMR.

References:

[1] Alexander, G.L.: Microgram XXVII, p.268, (1994)

Substance Number 16 in Chapter 2.7

IUPAC Name: 5-Phenyl-4,5-dihydro-oxazol-2-ylamine
Synonyms: 2-Amino-4,5-dihydro-5-phenyloxazole
4,5-Dihydro-5-phenyl-2-oxazolamine
Aminorex
Menocil

Chemical Structure:

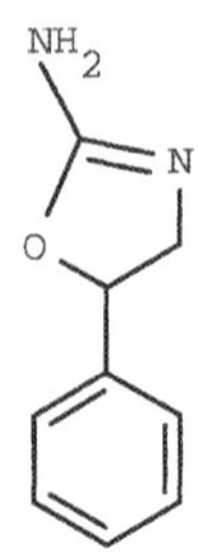

MF: C9H10N2O
MW: 162.191
CA Registry Number: [2207-50-3]
CA Chemical Name: Oxazole, 2-amino-4,5-dihydro-5-phenyl-
Category: CNS stimulant
Street Names: Aminorex
Abuse: Frequent
Type of action: Stimulant
Human active dose: 15-25 mg
Duration of action: Not reported
Toxic manifestations: Hyperactivity, confusion, hypertension, cardiovascular disturbances, hyperthermia, convulsions
Toxicity: Not reported

This analogue of 4-methylaminorex has been frequently seized in the USA [1,2]. These references also describe the identification of this substance by GC-MS and NMR. Also, analytical profiles (UV, IR, MS and HPLC) of aminorex and its positional isomer (« rexamino »; 2-amino-4,5-dihydro-4-phenyloxazole) have been described [3]. Originally, the compound was marketed as an anorectic (Menocil) but had to be withdrawn from the market because of high incidence of potentially fatal pulmonary hypertension [4].

References:

[1] Klein, R.F.X.; Morello, D.R.: Microgram, XXIII, p.119, (1990)
[2] Brewster, M.E.; Davis, F.T.: J. Forensic Sci., **36**, p.587, (1991)
[3] Noggle, F.T.; Clark, C.R.; DeRuiter, J.: Microgram, XXIII, p.281, (1990)
[4] Seiler, K.: Arzneimittel Forschung, **25**, p.837, (1975)

Substance Number 17 in Chapter 2.7

IUPAC Name: Diphenyl-2-pyrrolidin-2-yl-methanol
Synonyms: α,α-Diphenyl-(2-pyrrolidinyl)methanol
α,α-Diphenylprolinol
α,α-Diphenyl-(2-pyrrolidinyl)carbinol

Chemical Structure:

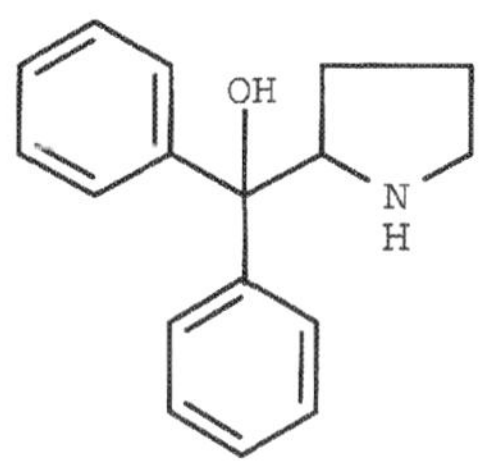

MF: C17H19NO
MW: 253.343
CA Registry Number: Racemate [112022-88-53]; [R]-[22348-32-9]; [S]-[112068-01-6]
CA Chemical Name: 2-Pyrrolidinemethanol, α,α-diphenyl-
Category: CNS Stimulant
Street Names: Not reported
Abuse: Rare
Type of action: Stimulant, euphoriant
Human active dose: 2-5 mg
Duration of action: Long
Toxic manifestations: Hyperactivity, confusion, hypertension, cardiovascular disturbances, hyperthermia, convulsions
Toxicity: Not reported

The CNS stimulant properties and very easy synthesis [1] of this compound were described by a sharing clandestine chemist on the Internet [2].

References:

[1] Winthrop, S.O.; Humber, L.G.: J. Org. Chem., **26**, p.2834, (1961)
[2] Newsgroups: Alt.drug.chemistry (1995)

Substance Number 18 in Chapter 2.7

IUPAC Name: 1-Phenyl-ethylamine
Synonyms: α-Methylbenzylamine
α-Phenethylamine

Chemical Structure:

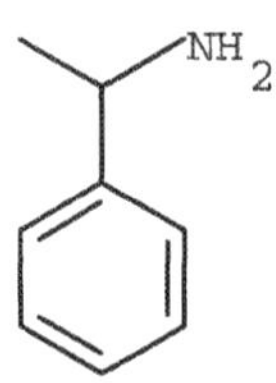

MF: C8H11N
MW: 121.182
CA Registry Number: [98-84-0]; racemate [618-36-0]; [R]-[3886-69-9]; [S]-[2627-86-3]
CA Chemical Name: Benzenemethanamine, α-methyl-
Category: CNS stimulant
Street Names: Not reported
Abuse: Frequent
Type of action: Stimulant (?)
Human active dose: Not reported
Duration of action: Not reported
Toxic manifestations: Not reported
Toxicity: 240 mg/kg, i.p. (mice).

Large quantities of this phenethylamine (racemic) have repeatedly been seized on the clandestine drug market in Holland [1]. The substance is only a weak CNS stimulant [2]. Isolated cases of seizure of its (+) isomer [1] and N-methyl derivative [3] have also been reported. This reference also describes the identification of this drug by MS and HPLC.

References:

[1] King. L.A.; Poortman-van der Meer, A.J.; Huizer, H.: Forensic Sci. Int., **77**, p.141, (1996)
[2] Van der Schoot, J.B.; Ariens, E.J.; Van Rossum, J.M.; Hurkmans, J.A.: Arzneim. Forschung, **9**, p.902, (1961)
[3] Clark, C.C.: Microgram, XXVI, p.90, (1993)

Substance Number 19 in Chapter 2.7

IUPAC Name: 1-p-Tolyl-ethylamine
Synonyms: α,4-Dimethylbenzylamine
4-Methyl-α-phenethylamine

Chemical Structure:

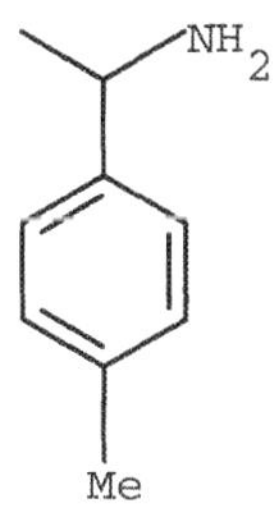

MF: C9H13N
MW: 135.208
CA Registry Number: [586-70-9]; [R]-[4187-38-6]; [S]-[27298-98-2]
CA Chemical Name: Benzenemethanamine, α,4-dimethyl-
Category: CNS stimulant
Street Names: Not reported
Abuse: Rare
Type of action: Stimulant (?)
Human active dose: Not reported
Duration of action: Not reported
Toxic manifestations: Not reported
Toxicity: Not reported

This substance has been seized on the clandestine drug market in Great Britain [1] and Australia [2]. The substance seems to be only a weak CNS stimulant [3]. The reference [1] also describes the identification of this drug by MS and HPLC.

References:

[1] Groombridge, C.J.; Hooker, R.: Microgram, XXIX, (2), p.38, (1996)
[2] Sims, D.N.; Kirkbride, K.P.: Microgram, XVIII, p.152, (1985)
[3] King. L.A.; Poortman-van der Meer, A.J.; Huizer, H.: Forensic Sci. Int., **77**, p.141, (1996)

2.8 Synthetic Opiates

This chapter is devoted to the designer drugs exhibiting morphine-like analgesic properties. In man, these drugs produce euphoria, sedation, altered pain perception, suppression of anxiety along with less desirable effects such as miosis, nausea, vomiting, respiratory depression, orthostatic hypotension, intestinal motility disturbances and convulsions. Pulmonary oedema is the most frequent and potentially lethal complication of the opiate overdose. Chronic opiate abuse leads to the development of tolerance and psychical and physical dependence. Clinical toxicology and management of opiate poisoning have been described [1].

This class of drugs is very important because a huge number of synthetic opiates have been synthesised in attempts to find an ideal potent analgesic free of habit forming properties. The number of these substances has been estimated to be about 4000 [2].

The term « designer drug » was coined also for one of these compounds which was detected in California in 1979 [3].

This « inverted » analogue (MPPP, **1**) of meperidine (**2**) is a powerful analgesic synthesised in 1947 [21] but for unknown reasons it has never been utilised in medicine. This compound, which may be also considered to be an analogue of another synthetic opiate alphaprodine (**3**), was « rediscovered » by two businessmen in an attempt to produce a very profitable drug and to circumvent the Controlled Substance Act [4].

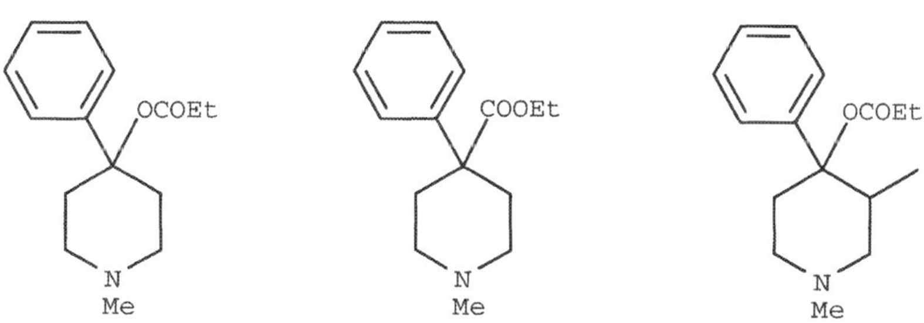

1 3-5 mg | **2** 50-100 mg | **3** 5-10 mg

Unfortunately, the produced drug contained a very toxic impurity (MTPT, **4**) because of rather harsh reaction conditions. MPTP was later found to destroy selectively dopamine neurones in the brain and produce symptoms of Parkinson's disease in users of the contaminated drug.

Consequently, the meperidine analogues have become more rare on the clandestine market. There are two powerful synthetic opiates (**5** and **6**) belonging to this group of designer drugs which have been seized in isolated cases.

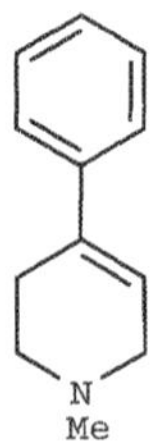

4

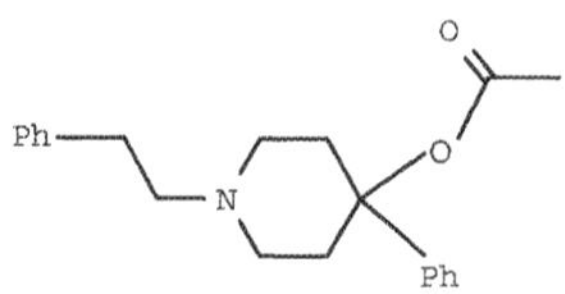

5

5-10 mg

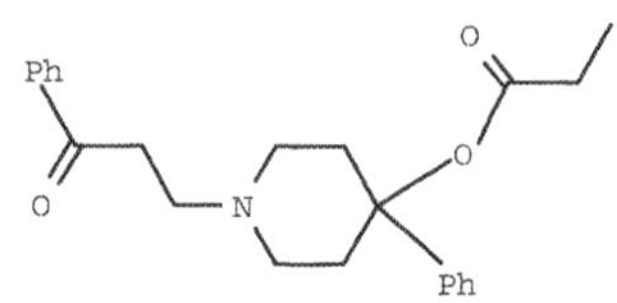

6

About 0.1 mg

Nevertheless, the meperidine analogues are still products of interest to a clandestine chemist because they are sufficiently potent and they can be easily manufactured from unrestricted and readily available chemicals.

Fentanyl (**7**) [5] and its analogues represent another group of analgesics which is particularly attractive for a clandestine manufacturer.

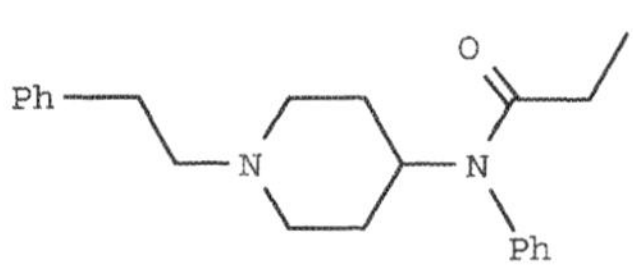

7

0.1 mg

Fentanyl is an extremely potent and rapidly acting analgesic which is currently utilised in medicine. However, its clandestine production is frequent because it is a very lucrative activity. Thus, a benefit of about $2,000,000 may be gained from investing only $200 for the chemical precursors [6].

Out of 1400 potential fentanyl analogues [6], more than 220 compounds [7] have been described in the literature [8,9,10].

Most of these substances are extremely powerful analgesics and their chemical structures allow a great number of variations, usually without a considerable loss of potency. These designer drugs appeared in 1979 [3] and their number was rapidly growing. Their presence in the clandestine market gave rise to many suspect deaths by overdose, but due to the high potency of fentanyls (i.e. very low plasma concentration) no drug was detected by routine analytical tests.

Thus, the first fentanyl analogue, α-methylfentanyl (« China White », « Synthetic », **8**) [11] appeared in 1979. It was followed by 4-fluorofentanyl (**9**) in 1981 and by fentanyl (**7**) of clandestine origin in 1982.

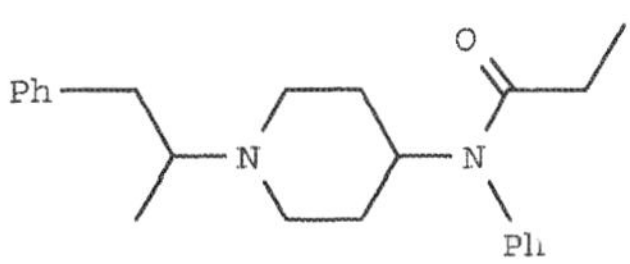

8
0.05-0.125 mg
(1979)

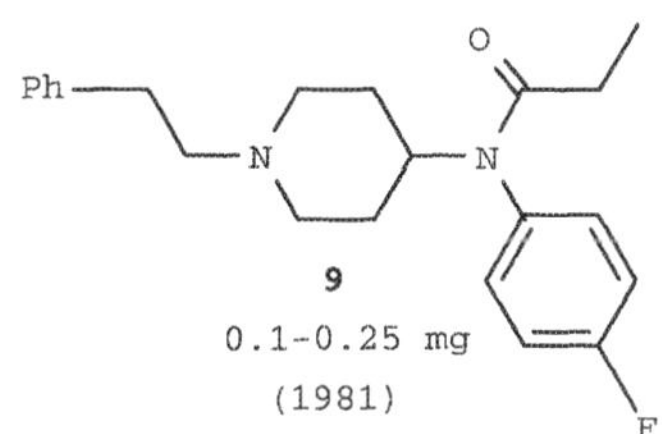

9
0.1-0.25 mg
(1981)

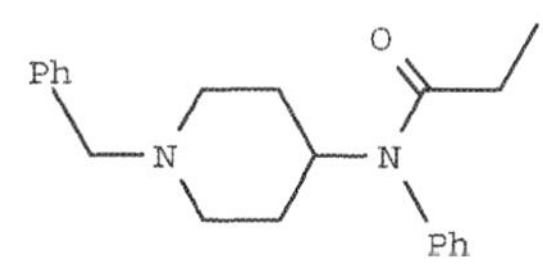

10
>10 mg
(1982)

Sixteen fentanyl analogues have recently appeared in illicit street opiate samples [3]. Human pharmacology of these illicit substances remains unknown or poorly investigated. Human active doses are known only for several substances and considerable differences of the reported values have been observed [12].

The analgesic potency of these substances is estimated to be 30 to 100 times higher than that of morphine, fentanyl itself being 100 times more potent.

Of these analogues, only benzylfentanyl (**10**) has been reported to be a less potent analgesic than morphine.

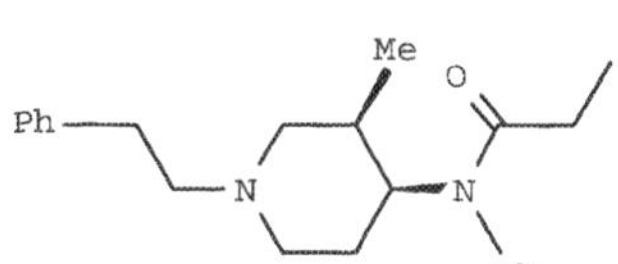

11
0.005-0.01 mg
(1983)

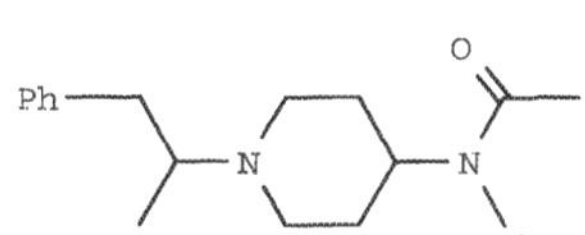

12
About 1 mg
(1983)

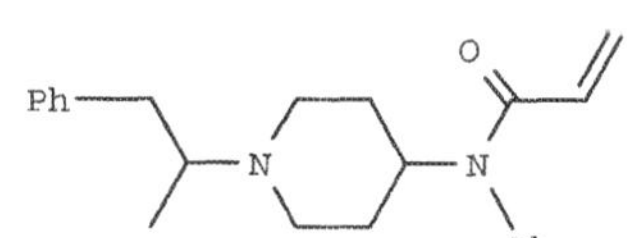

13
(1983)

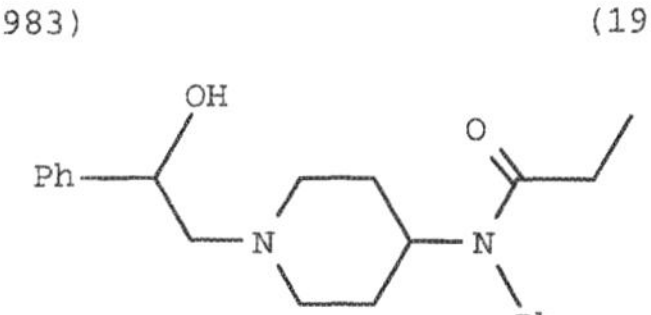

14
0.1-0.25 mg

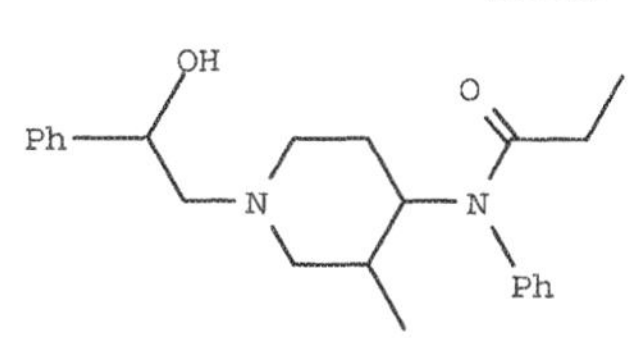

15
About 0.01 mg

However, cis-3-methylfentanyl (**11**) [13] is one of the most powerful analgesics with approximately 15 times the potency of fentanyl. In man 300 μg of this substance may lead to very serious respiratory depression and death. Also, its hydroxyderivative (ohmefentanyl, **15**) is an extremely potent opiate.

The heterocyclic fentanyl analogues (**16, 17, 18, 19, 20**) are also powerful but short acting analgesics [14].

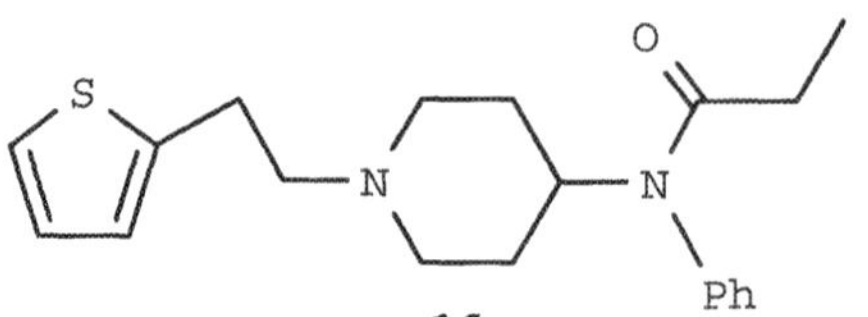

16
0.05-0.25 mg
(1985)

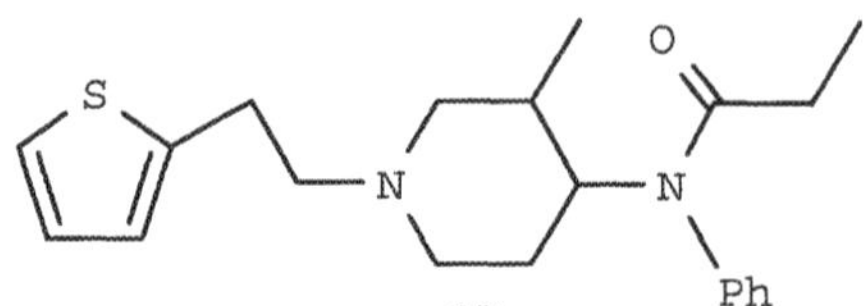

17
About 0.01 mg
(1985)

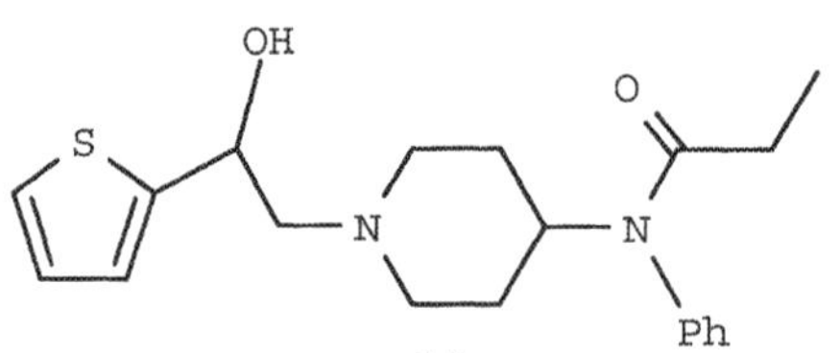

18
0.05-0.125 mg
(1985)

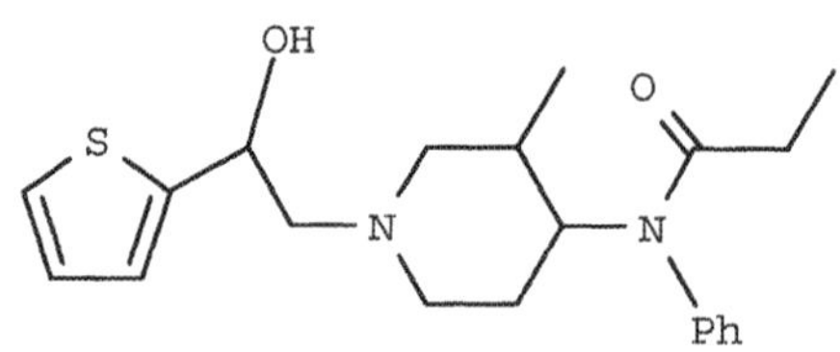

19
About 0.01 mg
(1985)

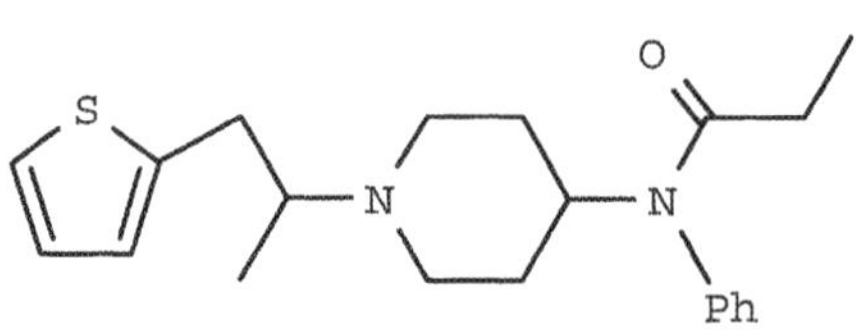

20
About 0.1 mg
(1985)

Interestingly, the fentanyls produce psychomotoric excitation in horse. Therefore, several illicit fentanyl analogues, particularly cis-3-methylfenanyl and α-fentanyl, have been used in racehorses doping [24].

The illicit fentanyls are often produced by professionally trained chemists [15]. Consequently, new, unknown and sophisticated fentanyl analogues, which may be very difficult to identify, are likely to appear in the future in illicit traffic.

The detection and identification of fentanyl and its analogues by GC and GC-MS have been described [16,17,18]. Another ingenious GC detection method is based on the fact that many fentanyl analogues are metabolised to the same normetabolite which can be easily derivatised [23].

Due to the very low plasmatic concentration of these drugs, the recently developed analytical techniques generally utilise sensitive immunoassay [19], or radioimmunoassay [20] methods. A GC procedure for determination of fentanyl in human hair has also been published [22].

The identification of new and unknown fentanyl analogues using selective extraction techniques and spectroscopic methods (IR, NMR, MS) has also been described [7]. In addition, an excellent, comprehensive study [25] reports all C13 chemical shifts for fentanyl and its fifteen analogues.

2.8.1 General References

[1] Easom, J.M.; Lovejoy, F.H.: Opiates. In: Clinical Management of Poisoning and Drug Overdose, Eds. Haddad, L.M.; Winchester, J.F.; W.B. Saunders Company, Eastbourne, UK, p.424, (1983)

[2] Burger, A.: In: Medicinal Chemistry, Wiley-Interscience Publishers, New York, p.1338, (1970)

[3] Henderson, G.: J. Forensic Sci., **33**, p.569, (1988)

[4] Weingarten, H.L.: J. Forensic Sci., **33**, p.588, (1988)

[5] Janssen, P.A.: Brit. J. Anaesthesia, **34**, p.260, (1962)

[6] LaBarbera, M.; Wolfe, T.: J. of Psychoactive Drugs, **15**, p.294, (1983)

[7] Cooper, D.; Jacob, M.; Allen, A.: J. Forensic Sci., **31**, p.511, (1986)

[8] Bagley, J.R.; Riley, T.N.: J. Med. Chem., **22**, p.1167, (1979)

[9] Van Bever, W.F.M.; Niemegeers, C.J.E.; Janssen, P.A.J.: J. Med. Chem., **17**, p.1047, (1974)

[10] Lurie, I.S.; Allen. A.C.: J. Chromatogr., **292**, p 289, (1984)

[11] Ayres, W.A.; Starsiak, M.J.; Sokolay, P.: J. of Psychoactive Drugs, **13**, p.91, (1981)

[12] Ref. 6, p.293

[13] Esposito, F.M.; Winek, C.L.: J. Forensic Sci., **36**, p.86, (1991)

[14] Poklis, A.: Clin. Toxicol., **33**, p.439, (1995)

[15] Chem. Eng. News, March 10, p.13, (1986)

[16] Van Rooy, H.H. et al.: J. Chromatogr., **223**, p.85, (1981)

[17] Moore, J.M.; Allen, A.C.; Cooper, D.A.; Carr, S.M.: Anal. Chem., **58**, p.1656, (1986)

[18] Ruangyuttikarn, W.; Law, M.Y.; Rollins, D.E.; Moody, D.E.: J. Anal. Toxicol., **14**, p.160, (1990)

[19] Schwartz, J.G. et al.: Am. J. Forensic Med. Pathol., **15**, p.236, (1994)

[20] Henderson, G.; Harkey, M.R.; Jones, A.D.: J. Anal. Toxicol., **14**, p.172, (1990)

[21] Ziering, A.; Berger, L.; Heineman, S.D.; Lee, J.: J. Org. Chem., **12**, p.898, (1947)

[22] Selavka, C.M.; Mason, A.P.; Riker, C.D.; Crookham, S.: J. Forensic Sci., **40**, p.681, (1995)

[23] Henderson, G.L.; Hammargren, W.R.: J. Analyt. Toxicol., **12**, p.183, (1988)

[24] Tobin, T.; Tai, H.H.; Tai, C.L.: Res. Comm. Pathol. Pharmacol., **60**, p.97, (1988)

[25] Brine, G.A.; Boldt, K.G.; Huang, P.T.; Sawyer, D.K.; Carroll F.I.: J. Heterocyclic Chem., **26**, p.677, (1989)

2.8.2. Data Sheets: Synthetic Opiates

Warning: Only few human data concerning these designer drugs can be found in the literature. Hence, the human doses given here are estimated from animal studies, similarly to Esposito, et al. (1991) and Cooper, et al. (1986) (see General References of this section), the analgesic potency of fentanyl being taken as 100 times that of morphine. This admittedly incorrect and inaccurate procedure should give the reader an idea about the relative potency of these designer analogues and avoid a complicated discussion of the available and often controversial animal data. Consequently, these values do not represent any accurate or experimental human data and must not be treated as such.

Substance Number 1 in Chapter 2.8

IUPAC Name: Propionic acid, 1-methyl-4-phenyl-piperidin-4-yl ester
Synonyms: 1-Methyl-4-phenyl-4-propionyloxy-piperidine
1-Methyl-4-phenyl-4-piperidyl propionate
MPPP

Chemical Structure:

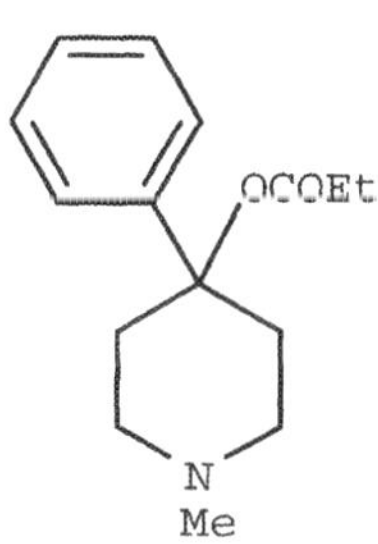

MF: C15H21NO2
MW: 261.363
CA Registry Number: [13147-09-6]
CA Chemical Name: 4-Piperidinol, 1-methyl-4-phenyl-, propanoate (ester)
Category: Synthetic opiate
Street Names: Synthetic heroin; New heroin
Abuse: Rare
Type of action: Euphoriant, analgesic
Human active dose: 3-5 mg
Duration of action: Not reported
Toxic manifestations: Miosis, nausea, vomiting, orthostatic hypotension, constipation, convulsions, muscular hyperactivity, respiratory depression, pulmonary oedema
Toxicity: Not reported

This analogue of both meperidine and alphaprodine is a potent synthetic analgesic of opiate type [1]. It was detected in California in 1979 [2,3,4]. These references also describe the identification of this drug by GC, MS and IR. The clandestinely manufactured drug frequently contained a neurotoxic impurity, causing Parkinson's disease in its users. Many people became irreversibly disabled and the drug practically disappeared from the market [2].

References:

[1] Ziering, A.; Berger, L.; Heineman, S.D.; Lee, J.: J. Org. Chem., **12**, p.898, (1947)
[2] Weingarten, H.L.: J. Forensic Sci., **33**, p.588, (1988)
[3] Heagy, J.A.: Microgram, XV, p.181, (1982)
[4] Jindal, S.P.; Lutz, T.; Bagchi, S.P.: J. Chromatogr., **408**, p.356, (1987)

Substance Number 5 in Chapter 2.8

IUPAC Name: Acetic acid, 1-phenethyl-4-phenyl-piperidin-4-yl (ester)
Synonyms: 1-Phenethyl-4-phenyl-4-piperidinol acetate (ester)
1-(2-Phenethyl)-4-phenyl-4-acetyloxypiperidine
PEPAP

Chemical Structure:

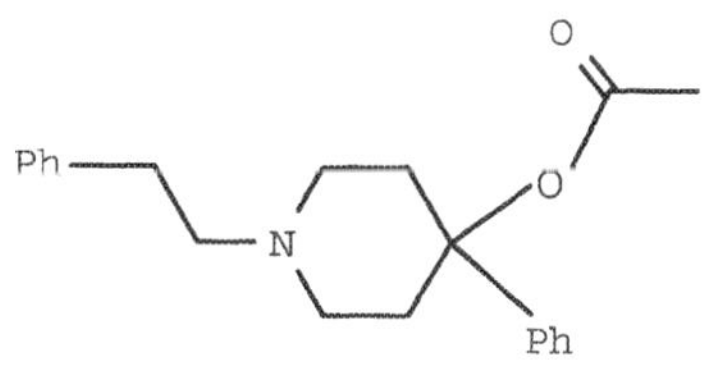

MF: C21H25NO2
MW: 323.433
CA Registry Number: [64-52-8]
CA Chemical Name: 4-Piperidinol, 4-phenyl-1-(2-phenylethyl)-, acetate (ester)
Category: Synthetic opiate
Street Names: Synthetic heroin; New heroin
Abuse: Limited
Type of action: Euphoriant, analgesic
Human active dose: 5-10 mg
Duration of action: Not reported
Toxic manifestations: Miosis, nausea, vomiting, orthostatic hypotension, constipation, convulsions, muscular hyperactivity, respiratory depression, pulmonary oedema
Toxicity: Not reported

This meperidine analogue was seized several times in the USA and Canada [1] in the 1980s. Determination of this substance by GC-ECD has been described [2]. The substance has been reported to be roughly as potent an analgesic as morphine [3].

References:

[1] Microgram, XXII, (4), p.52, (1989)
[2] Moore, J.M.; Allen, A.C.; Cooper, D.A.; Carr, S.M.: Anal. Chem., **58**, p.1656, (1986)
[3] Casy, A.F.; Armstrong, N.A.: J. Med. Chem., **8**, p.57, (1965)

Substance Number 6 in Chapter 2.8

IUPAC Name: Propionic acid, 1-(3-oxo-3-phenyl-propyl)-4-phenyl-piperidin-4-yl ester
Synonyms: 1-(3-Oxo-3-phenylpropyl)-4-phenyl-4-piperidinol propionate
OPPPP

Chemical Structure:

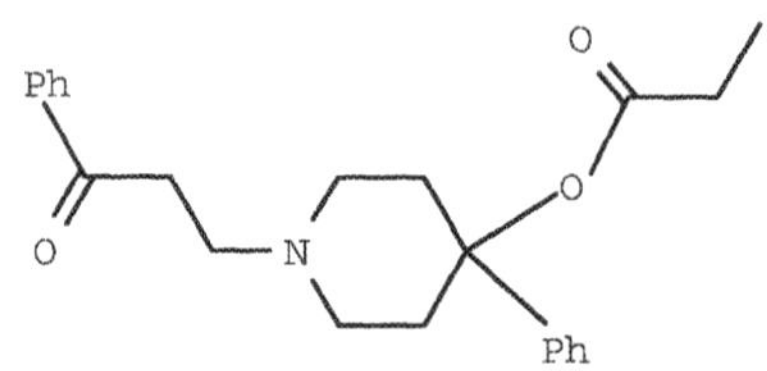

MF: C23H27NO3
MW: 365.470
CA Registry Number: Not reported
CA Chemical Name: 4-Piperidinol, 1-(3-oxo-3-phenylpropyl)-4-phenyl-, propanoate (ester)
Category: Synthetic opiate
Street Names: Synthetic heroin; New heroin
Abuse: Limited
Type of action: Euphoriant, analgesic
Human active dose: About 0.1 mg
Duration of action: Not reported
Toxic manifestations: Miosis, nausea, vomiting, orthostatic hypotension, constipation, convulsions, muscular hyperactivity, respiratory depression, pulmonary oedema
Toxicity: Not reported

This powerful meperidine analogue was seized several times in the USA and Canada [1] in the 1980s. The compound has been reported to be 1000 times more potent than meperidine [2].

References:

[1] Microgram, XXII, (4), p.52, (1989)
[2] Carabateas, P.M.; Grumbach, L.: J. Med. Pharm. Chem., **5**, p.913, (1962)

Substance Number 8 in Chapter 2.8

IUPAC Name: N-[1-(1-Methyl-2-phenyl-ethyl)-piperidin-4-yl]-N-phenyl-propionamide
Synonyms: N-[1-(2-Phenylisopropyl)-4-piperidyl]-N-phenylpropionamide
α-Methyfentanyl
AMF

Chemical Structure:

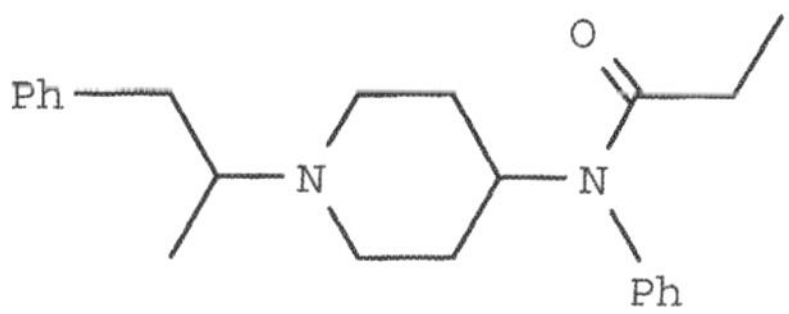

MF: C23H30N2O
MW: 350.503
CA Registry Number: [79704-88-4]; hydrochloride [1443-44-3]
CA Chemical Name: Propanamide, N-[1-(1-methyl-2-phenylethyl)-4-piperidinyl]-N-phenyl-
Category: Synthetic opiate
Street Names: China White; Persian White; Mexican Brown; China town; Poison; Tango & Cash; Friend; Goodfellas, Synthetic
Abuse: Limited
Type of action: Euphoriant, analgesic
Human active dose: 0.05-0.125 mg
Duration of action: Short
Toxic manifestations: Miosis, nausea, vomiting, orthostatic hypotension, constipation, convulsions, muscular hyperactivity, respiratory depression, pulmonary oedema
Toxicity:. Not reported; LD50 for fentanyl is 11.2 mg/kg i.v. (mouse).

The drug was detected in California in 1979 [1]. Originally, it was sold as « China White » but it may appear under the street names common for all fentanyl analogues. The substance is reported to be about twice as potent as fentanyl [2]. Its identification by GC-ECD [3] (including its determination), GC, IR, MS, NMR [4,8], C13-NMR [9] and by HPLC [5] has been described. In the enzyme linked immunosorbent assay (ELISA) the compound shows rather poor cross-reactivity with fentanyl [6] but in the solid phase radio-immunoassay screening method it gives good response [7].

References:

[1] Henderson, G.: J. Forensic Sci., **33**, p.569, (1988)
[2] Ref. 1, p.570
[3] Moore, J.M.; Allen, A.C.; Cooper, D.A.; Carr, S.M.: Anal. Chem., **58**, p.1656, (1986)

[4] Cooper, D.; Jacob, M.; Allen, A.: J. Forensic Sci., **31**, p.511, (1986)
[5] Lurie, I.S.; Allen. A.C.: J. Chromatogr., **292**, p.289, (1984)
[6] Ruangyuttikarn, W.; Law, M.Y.; Rollins, D.E.; Moody, D.E.: J. Anal. Toxicol., **14**, p.160, (1990)
[7] Henderson, G.; Harkey, M.R.; Jones, A.D.: J. Anal. Toxicol., **14**, p.172, (1990)
[8] Mills III, T.; Roberson, J.C.; McCurdy, H.H.; Wall, W.H.: Instrumental Data for Drug Analysis, Elsevier Science Publishing Co., Inc., p.1472, (1992)
[9] Brine, G.A.; Boldt, K.G.; Huang, P.T.; Sawyer, D.K.; Carroll, F.I.: J. Heterocyclic Chem., **26**, p.677, (1989)

Substance Number 9 in Chapter 2.8

IUPAC Name: N-(4-Fluoro-phenyl)-N-(1-phenethyl)-piperidin-4-yl)-propionamide
Synonyms: N-(4-Fluorophenyl)-N-[1-(2-phenethyl)-4-iperidyl]-propionamide
4-Fluorofentanyl

Chemical Structure:

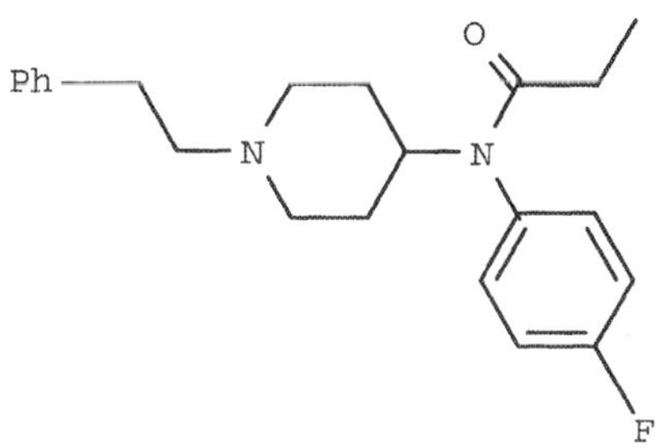

MF: $C_{22}H_{27}FN_2O$
MW: 354.467
CA Registry Number: [90736-23-5]; hydrochloride [117332-92-0]
CA Chemical Name: Propanamide, N-(4-fluorophenyl)-N-[1-(2-phenylethyl)-4-piperidinyl]-
Category: Synthetic opiate
Street Names: China White; Persian White; Mexican Brown; China town; Poison; Tango & Cash; Friend; Goodfellas
Abuse: Limited
Type of action: Euphoriant, analgesic
Human active dose: 0.1-0.25 mg
Duration of action: Not reported
Toxic manifestations: Miosis, nausea, vomiting, orthostatic hypotension, constipation, convulsions, muscular hyperactivity, respiratory depression, pulmonary oedema
Toxicity: Not reported

The drug was detected in California in 1981 [1]. It is usually sold under the street names common for all fentanyl analogues. The substance is reported to be about as potent as fentanyl [2]. Its identification by GC-ECD [3] (including its determination), GC, IR, MS and NMR [4,] and by HPLC [5] has been described. In the enzyme linked immunosorbent assay (ELISA) the compound shows good cross-reactivity with fentanyl [6]. Also, in the solid phase radioimmunoassay screening method it gives good response [7].

References:

[1] Henderson, G.: J. Forensic Sci. **33**, p.569, (1988)
[2] Ref. 1, p.572

[3] Moore, J.M.; Allen, A.C.; Cooper, D.A.; Carr, S.M.: Anal. Chem., **58**, p.1656, (1986)
[4] Cooper, D.; Jacob, M.; Allen, A.: J. Forensic Sci., **31**, p.511, (1986)
[5] Lurie, I.S.; Allen. A.C.: J. Chromatogr., **292**, p 289, (1984)
[6] Ruangyuttikarn, W.; Law, M.Y.; Rollins, D.E.; Moody, D.E.: J. Anal. Toxicol., **14**, p.160, (1990)
[7] Henderson, G.; Harkey, M.R.; Jones, A.D.: J. Anal. Toxicol., **14**, p.172, (1990)

Substance Number 10 in Chapter 2.8

IUPAC Name: N-[1-Benzyl-piperidin-4-yl]-N-phenyl-propionamide
Synonyms: N-[1-Benzyl-4-piperidyl]-N-phenylpropionamide
Benzylfentanyl

Chemical Structure:

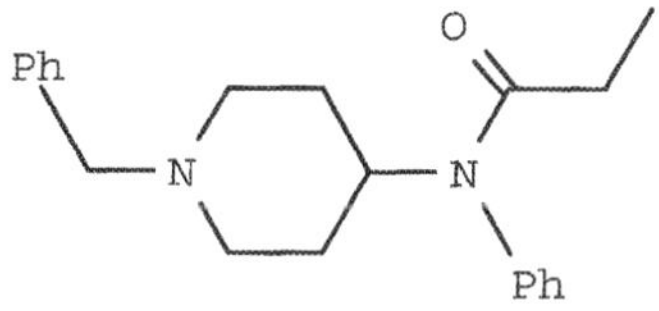

MF: C21H26N2O
MW: 322.449
CA Registry Number: [1474-02-8]; hydrochloride [5156-58-1]
CA Chemical Name: Propanamide, N-phenyl-N-[1-(phenylmethyl)-4-piperidinyl]-
Category: Synthetic opiate
Street Names: China White; Persian White; Mexican Brown; China town; Poison; Tango & Cash; Friend; Goodfellas
Abuse: Limited
Type of action: Euphoriant, analgesic
Human active dose: Above 10 mg
Duration of action: Not reported
Toxic manifestations: Miosis, nausea, vomiting, orthostatic hypotension, constipation, convulsions, muscular hyperactivity, respiratory depression, pulmonary oedema
Toxicity: Not reported

The drug was detected in California in 1981 [1]. It may appear under the street names common for all fentanyl analogues. The substance is reported to have low analgesic potency and is unable to replace morphine in dependent monkeys [2]. Its identification by GC-ECD [3] (including its determination), GC, IR, MS, NMR [4], C13-NMR [7] and by HPLC has been described [5]. In the enzyme linked immunosorbent assay (ELISA) the compound does not show any cross-reactivity with fentanyl [6].

References:

[1] Henderson, G.: J. Forensic Sci., **33**, p.569, (1988)
[2] Jacobson, A.: Biological Evaluation of Compounds for Their Physical Dependence and Abuse Liability XI. In: NIDA Res. Monogr., **81**, p.466, (pp.480-1), (1988)
[3] Moore, J.M.; Allen, A.C.; Cooper, D.A.; Carr, S.M.: Anal. Chem., **58**, p.1656, (1986)
[4] Cooper, D.; Jacob, M.; Allen, A.: J. Forensic Sci., **31**, p.511, (1986)

[5] Lurie, I.S.; Allen. A.C.: J. Chromatogr., **292**, p 289, (1984)
[6] Ruangyuttikarn, W.; Law, M.Y.; Rollins, D.E.; Moody, D.E.: J. Anal. Toxicol., **14**, p.160, (1990)
[7] Brine, G.A.; Boldt, K.G.; Huang, P.T.; Sawyer, D.K.; Carroll, F.I.: J. Heterocyclic Chem., **26**, p.677, (1989)

Substance Number 11 in Chapter 2.8

IUPAC Name: N-(3-Methyl-1-phenylethyl-piperidin-4-yl)-N-phenyl-propionamide
Synonyms: N-[3-Methyl-1-(2-phenylethyl)-4-piperidyl]-N-phenylpropionamide
cis-3-Metylfentanyl

Chemical Structure:

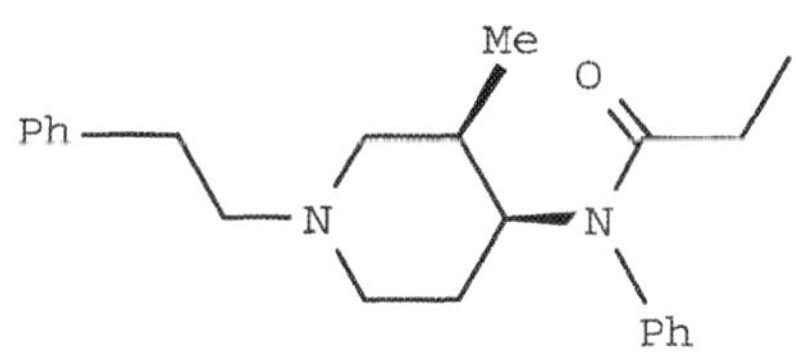

MF: C23H30N2O
MW: 350.503
CA Registry Number: [42045-86-3]; cis(-) [78995-19-4]; cis, racemate [53758-16-0]; cis(+) [53758-22-8]; trans, racemate [53758-18-2]; hydrochloride, cis, racemate [112239-82-4]
CA Chemical Name: Propanamide, N-[3-methyl-1-(2-phenylethyl)-4-piperidinyl]-N-phenyl-
Category: Synthetic opiate
Street Names: Super-heroin; China White; Persian White; Mexican Brown; China town; Poison; Tango & Cash; Friend; Goodfellas
Abuse: Frequent
Type of action: Euphoriant, analgesic
Human active dose: 0.005-0.01 mg
Duration of action: About 4 hours
Toxic manifestations: Miosis, nausea, vomiting, orthostatic hypotension, constipation, convulsions, muscular hyperactivity, respiratory depression, pulmonary oedema
Toxicity: In man, an estimated fatal dose of this substance is only about 300 μg [4].

The drug was detected in California in 1983 [1]. It may appear under the street names common for all fentanyl analogues. The drug found in illicit traffic is mostly a mixture of *cis* and *trans* isomers in a ratio of about 2 to 3:1 [8]. The pure substance (*cis*-isomer) is reported to be about 15 times as potent as fentanyl [2]. Its identification by GC-ECD [3] (including its determination), GC, IR, MS and NMR [4], GC-MS [8], and by HPLC [5] has been described. In the enzyme linked immunosorbent assay (ELISA) the compound shows rather poor cross-reactivity with fentanyl [6] but in the solid phase radioimmunoassay screening method it gives good response [7].

References:

[1] Henderson, G.: J. Forensic Sci., **33**, p.569, (1988)
[2] Van Bever, W.F.M.; Niemegeers, C.J.E.; Janssen, P.A.J.: J. Med. Chem., **17**, p.1047, (1974)
[3] Moore, J.M.; Allen, A.C.; Cooper, D.A.; Carr, S.M.: Anal. Chem., **58**, p.1656, (1986)
[4] Cooper, D.; Jacob, M.; Allen, A.: J. Forensic Sci., **31**, p.511, (1986)
[5] Lurie, I.S.; Allen. A.C.: J. Chromatogr., **292**, p 289, (1984)
[6] Ruangyuttikarn, W.; Law, M.Y.; Rollins, D.E.; Moody, D.E.: J. Anal. Toxicol., **14,** p.160, (1990)
|7] Henderson, G.; Harkey, M.R.; Jones, A.D.: J. Anal. Toxicol., **14**, p.172, (1990)
[8] Esposito, F.M.; Winek, C.L.: J. Forensic Sci., **36**, p.86, (1991)

Substance Number 12 in Chapter 2.8

IUPAC Name: N-[1-(1-Methyl-2-phenyl-ethyl)-piperidin-4-yl]-N-phenyl-acetamide
Synonyms: N-[1-(2-Phenylisopropyl)-4-piperidyl]-N-phenylacetamide
Acetyl-α-methyfentanyl

Chemical Structure:

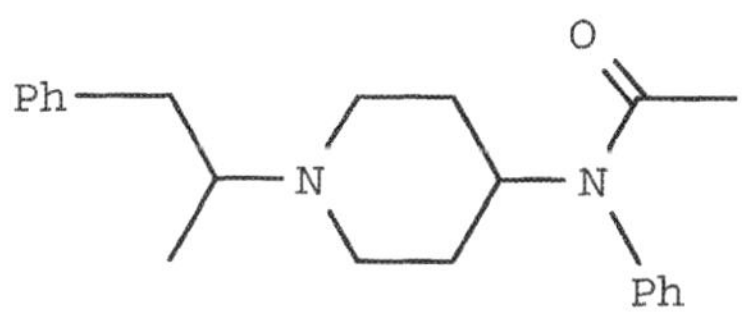

MF: C22H28N2O
MW: 336.46
CA Registry Number: [101860-00-8]
CA Chemical Name: Acetamide, N-[1-(1-methyl-2-phenylethyl)-4-piperidinyl]-N-phenyl-
Category: Synthetic opiate
Street Names: China White; Persian White; Mexican Brown; China town; Poison; Tango & Cash; Friend; Goodfellas
Abuse: Limited
Type of action: Euphoriant, analgesic
Human active dose: About 1 mg
Duration of action: Moderate
Toxic manifestations: Miosis, nausea, vomiting, orthostatic hypotension, constipation, convulsions, muscular hyperactivity, respiratory depression, pulmonary oedema
Toxicity: Not reported

The drug was detected in California in 1983 [1]. The substance may appear under the street names common for all fentanyl analogues. The compound is reported to have about 1/10 of fentanyl analgesic activity [2]. Its identification by GC-ECD [3] (including its determination), GC, IR, MS, NMR [4] has been described.

References:

[1] Henderson, G.: J. Forensic Sci., **33**, p.569, (1988)
[2] Jacobson, A.: Biological Evaluation of Compounds for Their Physical Dependence and Abuse Liability IX. In: NIDA Res. Monogr., **67**, p.397, (1986)
[3] Moore, J.M.; Allen, A.C.; Cooper, D.A.; Carr, S.M.: Anal. Chem., **58**, p.1656, (1986)
[4] Cooper, D.; Jacob, M.; Allen, A.: J. Forensic Sci., **31**, p.511, (1986)

Substance Number 13 in Chapter 2.8

IUPAC Name: N-[1-(1-Methyl-2-phenyl-ethyl)-piperidin-4-yl]-N-phenyl-acrylamide
Synonyms: N-[1-(2-Phenylisopropyl)-4-piperidyl]-N-phenyl-2-propenamide
Acryl-α-methylfentanyl

Chemical Structure:

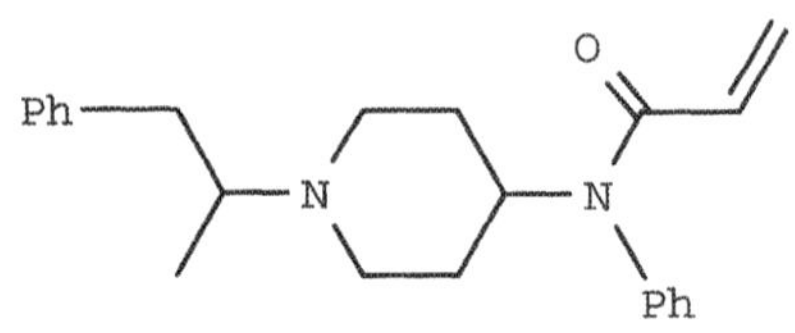

MF: C23H28N2O
MW: 348.487
CA Registry Number: [79279-03-1]
CA Chemical Name: 2-Propenamide, N-[1-(1-methyl-2-phenylethyl)-4-piperidinyl]-N-phenyl-
Category: Synthetic opiate
Street Names: China White; Persian White; Mexican Brown; China town; Poison; Tango & Cash; Friend; Goodfellas
Abuse: Limited
Type of action: Euphoriant, analgesic
Human active dose: Not reported
Duration of action: Not reported
Toxic manifestations: Miosis, nausea, vomiting, orthostatic hypotension, constipation, convulsions, muscular hyperactivity, respiratory depression, pulmonary oedema
Toxicity: Not reported

The drug was detected in California in 1983 [1]. This fentanyl analogue appeared as an impurity in street samples of α-methylacetylfentanyl [1]. Its identification by GC, IR, MS and NMR [2] has been described.

References:

[1] Henderson, G.: J. Forensic Sci., **33**, p.569, (1988)
[2] Cooper, D.; Jacob, M.; Allen, A.: J. Forensic Sci., **31**, p.511, (1986)

Substance Number 14 in Chapter 2.8

IUPAC Name: N-[1-(2-Hydroxy-2-phenyl-ethyl)-piperidin-4-yl]-N-phenyl-propionamide
Synonyms: N-[1-(2-Hydroxy-2-phenylethyl)-4-piperidyl]propionanilide
β-Hydroxyfentanyl

Chemical Structure:

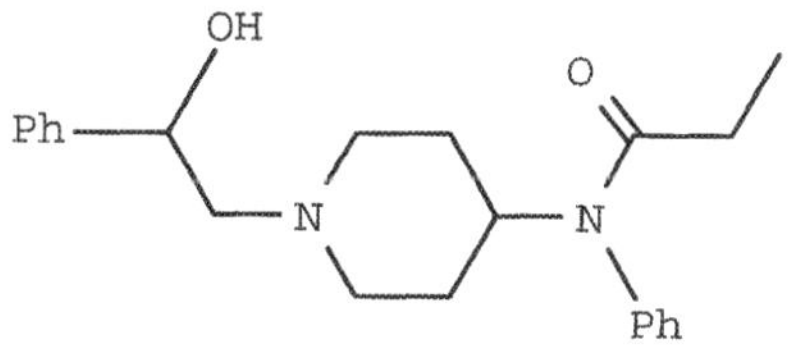

MF: C22H28N2O2
MW: 350.487
CA Registry Number: [78995-10-5]; hydrochloride [1473-95-6]
CA Chemical Name: Propanamide, N-[1-(2-hydroxy-2-phenylethyl)-4-piperidinyl]-N-phenyl-
Category: Synthetic opiate
Street Names: China White; Persian White; Mexican Brown; China town; Poison; Tango & Cash; Friend; Goodfellas
Abuse: Limited
Type of action: Euphoriant, analgesic
Human active dose: 0.1-0.25 mg
Duration of action: About 2 hours
Toxic manifestations: Miosis, nausea, vomiting, orthostatic hypotension, constipation, convulsions, muscular hyperactivity, respiratory depression, pulmonary oedema
Toxicity: Not reported

The drug was also detected in the illicit marketplace [1]. It may be sold under the street names common for all fentanyl analogues. Its identification by C13-NMR [1], MS [3] and an assay by RIA [2] have been described.

References:

[1] Brine, G.A.; Boldt, K.G.; Huang, P.T.; Sawyer, D.K.; Carroll, F.I.: J. Heterocyclic Chem., **26**, p.677, (1989)
[2] Watts, V.W.; Caplan, Y.H.: J. Anal. Toxicol., **14**, p.266, (1990)
[3] WHO Informational Manual on Designer Drugs, WHO/PSA/90.5, Geneva, p.21, (1991)

Substance Number 15 in Chapter 2.8

IUPAC Name: N-[1-(2-Hydroxy-2-phenyl-ethyl)-3-methyl-piperidin-4-yl]-N-phenyl-propionamide
Synonyms: N-[1-(2-Hydroxy-2-phenylethyl)-3-methyl-4-piperidyl]-propionanilide
β-Hydroxy-3-methylfentanyl
Ohmefentanyl

Chemical Structure:

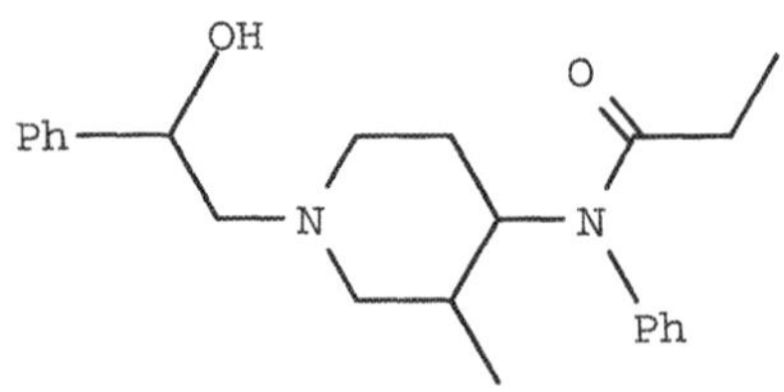

MF: C23H30N2O2
MW: 366.503
CA Registry Number: [78995-14-9]; hydrochloride [131724-54-4] and [135159-44-3]
CA Chemical Name: Propanamide, N-[1-(2-hydroxy-2-phenylethyl)-3-methyl-4-piperidinyl]-N-phenyl-
Category: Synthetic opiate
Street Names: China White; Persian White; Mexican Brown; China town; Poison; Tango & Cash; Friend; Goodfellas
Abuse: Limited
Type of action: Euphoriant, analgesic
Human active dose: About 0.01 mg
Duration of action: Not reported
Toxic manifestations: Miosis, nausea, vomiting, orthostatic hypotension, constipation, convulsions, muscular hyperactivity, respiratory depression, pulmonary oedema
Toxicity:. Not reported

This extremely potent opiate was also detected in the USA [1]. The cis-methyl isomer of this compound is in animal studies 28 times more potent than fentanyl [2]. One of its enantiomers (2S,3R,4S) is one of the most powerful opiates known [3].

The substance may appear under the street names common for all fentanyl analogues. Its HPLC separation [4] and MS spectrum [1] have been described.

References:

[1] WHO Informational Manual on Designer Drugs, WHO/PSA/90.5, Geneva, p.22, (1991)
[2] Jin, W.Q. et al.: Sci-Sin., **24**, p.710, (1981)
[3] Brine, G.A. et al.: J. Med. Chem., **38**, p.1547, (1995)
[4] Zhu, Y.C.; Prenant, C.; Crouzel, C.; Comar, D.; Chi, Z.Q.: J. Labelled Compd. Radiopharm., **31**, p.853, (1992)

Substance Number 16 in Chapter 2.8

IUPAC Name: N-Phenyl-N-[1-(2-thiophen-2-yl-ethyl)-piperidin-4-yl]-propionamide
Synonyms: N-Phenyl-N-[1-[2-(2-thienyl)ethyl]-4-piperidinyl]-propionamide
Thienylfentanyl

Chemical Structure:

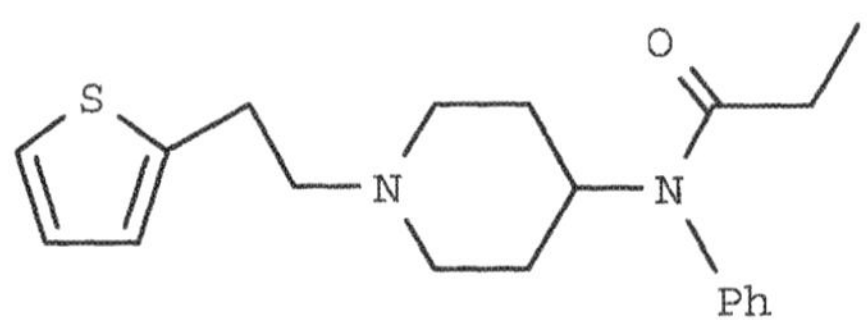

MF: C20H26N2OS
MW: 342.498
CA Registry Number: [1165-22-6]; hydrochloride [79278-88-9]
CA Chemical Name: Propanamide, N-phenyl- N-[1-[2-(2-thienyl)ethyl]-4-piperidinyl]-
Category: Synthetic opiate
Street Names: China White; Persian White; Mexican Brown; China town; Poison; Tango & Cash; Friend; Goodfellas
Abuse: Limited
Type of action: Euphoriant, analgesic
Human active dose: 0.05-0.25 mg
Duration of action: Short
Toxic manifestations: Miosis, nausea, vomiting, orthostatic hypotension, constipation, convulsions, muscular hyperactivity, respiratory depression, pulmonary oedema
Toxicity: Not reported

The drug was detected in California in 1985 [1]. It may be sold under the street names common for all fentanyl analogues. The substance is reported to be about as potent as fentanyl [2]. Its identification by GC, IR, MS, NMR [3] and C13-NMR [4] has been described. The substance can be detected by the solid phase radioimmunoassay screening method [5]. Its assay in urine samples by the RIA based method has also been reported [6].

References:

[1] Henderson, G.: J. Forensic Sci., **33**, p.569, (1988)
[2] Jacobson, A.: Biological Evaluation of Compounds for Their Physical Dependence and Abuse Liability XI. In: NIDA Res. Monogr., **81**, p.466 (pp.480-1), (1988)
[3] Cooper, D.; Jacob, M.; Allen, A.: J. Forensic Sci., **31**, p.511, (1986)

[4] Brine, G.A.; Boldt, K.G.; Huang, P.T.; Sawyer, D.K.; Carroll, F.I.: J. Heterocyclic Chem., **26**, p.677, (1989)
[5] Henderson, G.; Harkey, M.R.; Jones, A.D.: J. Anal. Toxicol., **14**, p.172, (1990)
[6] Watts, V.W.; Caplan, Y.H.: J. Anal. Toxicol., **14**, p.266, (1990)

Substance Number 17 in Chapter 2.8

IUPAC Name: N-[3-Methyl-1-(2-thiophen-2-yl-ethyl)-4-piperidin-4-yl]-N-phenyl-propionamide
Synonyms: N-[3-Methyl-1-[2-(2-thienyl)ethyl]-4-piperidinyl]-N-phenylpropionamide
cis-3-Methylthienylfentanyl

Chemical Structure:

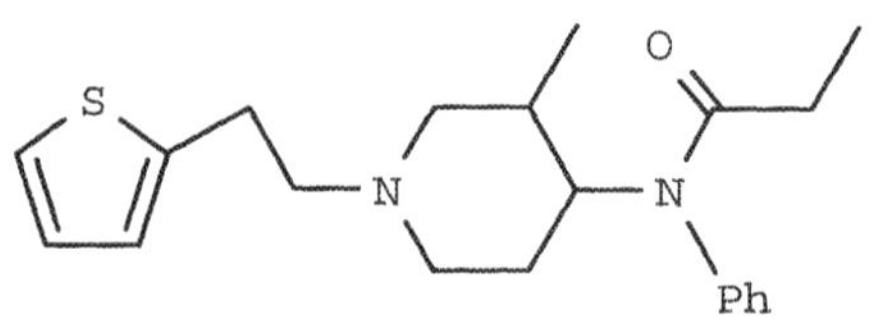

MF: C21H28N2OS
MW: 356.525
CA Registry Number: [86052-04-02]; hydrochloride [131690-45-4]
CA Chemical Name: Propanamide, N-[3-methyl-1-[2-(2-thienyl)ethyl]-4-piperidinyl]-N-phenyl-
Category: Synthetic opiate
Street Names: China White; Persian White; Mexican Brown; China town; Poison; Tango & Cash; Friend; Goodfellas
Abuse: Limited
Type of action: Euphoriant, analgesic
Human active dose: About 0.01 mg
Duration of action: Not reported
Toxic manifestations: Miosis, nausea, vomiting, orthostatic hypotension, constipation, convulsions, muscular hyperactivity, respiratory depression, pulmonary oedema
Toxicity: Not reported

The drug was detected in California in 1983 [1]. It may be sold under the street names common for all fentanyl analogues. The drug found in illicit traffic is mostly a mixture of *cis* and *trans* isomers. Its detection by GC-ECD [2] and its MS spectrum [3] have been described.

References:

[1] Henderson, G.: J. Forensic Sci., **33**, p.569, (1988)
[2] Henderson, G.L.; Hammargren, W.R.: J. Analyt. Toxicol., **12**, p.183, (1988)
[3] WHO Informational Manual on Designer Drugs, WHO/PSA/90.5, Geneva, p.18, (1991)

Substance Number 18 in Chapter 2.8

IUPAC Name: N-[1-(2-Hydroxy-2-thiophen-2-yl-ethyl)-piperidin-4-yl]-N-phenyl-propionamide
Synonyms: N-Phenyl-N-[1-[2-hydroxy-2-(2-thienyl)ethyl]-4-piperidinyl]propionamide
β-Hydroxythienylfentanyl

Chemical Structure:

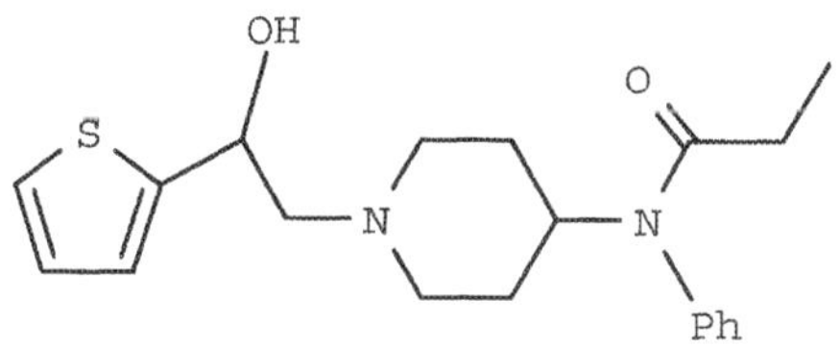

MF: C20H26N2O2S
MW: 358.497
CA Registry Number: [1474-34-6]
CA Chemical Name: Propanamide, N-[1-[2-hydroxy-2-(2-thienyl)ethyl]-4-piperidinyl]-N-phenyl-
Category: Synthetic opiate
Street Names: China White; Persian White; Mexican Brown; China town; Poison; Tango & Cash; Friend; Goodfellas
Abuse: Limited
Type of action: Euphoriant, analgesic
Human active dose: 0.05-0.125 mg
Duration of action: Not reported
Toxic manifestations: Miosis, nausea, vomiting, orthostatic hypotension, constipation, convulsions, muscular hyperactivity, respiratory depression, pulmonary oedema
Toxicity: Not reported

The drug was detected in California in 1985 [1]. It may be sold under the street names common for all fentanyl analogues. Its detection by GC-ECD, MS [2] and its C13-NMR spectrum [3] have been described.

References:

[1] Henderson, G.: J. Forensic Sci., **33**, p.569, (1988)
[2] Henderson, G.L.; Hammargren, W.R.: J. Analyt. Toxicol., **12**, p.183, (1988)
[3] Brine, G.A.; Boldt, K.G.; Huang, P.T.; Sawyer, D.K.; Carroll, F.I.: J. Heterocyclic Chem., **26**, p.677, (1989)

Substance Number 19 in Chapter 2.8

IUPAC Name: N-[1-(2-Hydroxy-2-thiophen-2-yl-ethyl)-3-methyl-piperidin-4-yl]-N-phenyl-propionamide
Synonyms: N-[1-[2-Hydroxy-2-(2-thienyl)ethyl]-3-methyl-4-piperidinyl]-N-phenyl]propionamide
β-Hydroxy-3-methylthienylfentanyl

Chemical Structure:

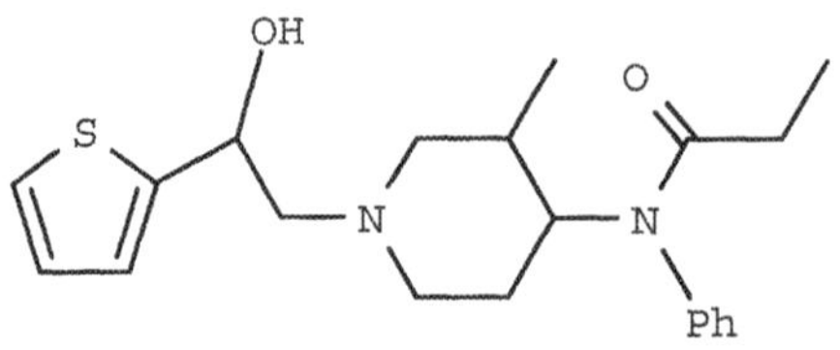

MF: C21H28N2O2S
MW: 372.524
CA Registry Number: [86052-03-1]
CA Chemical Name: Propanamide, N-[-1-[2-hydroxy-2-(2-thienyl)ethyl]-3-methyl-4-piperidinyl]-N-phenyl-
Category: Synthetic opiate
Street Names: China White; Persian White; Mexican Brown; China town; Poison; Tango & Cash; Friend; Goodfellas
Abuse: Limited
Type of action: Euphoriant, analgesic
Human active dose: About 0.01 mg
Duration of action: Not reported
Toxic manifestations: Miosis, nausea, vomiting, orthostatic hypotension, constipation, convulsions, muscular hyperactivity, respiratory depression, pulmonary oedema
Toxicity: Not reported

The drug was detected in California in 1985 [1]. It may be sold under the street names common for all fentanyl analogues. The drug found in illicit traffic is mostly a mixture of *cis* and *trans* isomers. Its detection by GC-ECD and MS [2] has been described.

References:

[1] Henderson, G.: J. Forensic Sci., **33**, p.569, (1988)
[2] Henderson, G.L.; Hammargren, W.R.: J. Analyt. Toxicol., **12**, p.183, (1988)

Substance Number 20 in Chapter 2.8

IUPAC Name: N-[1-(1-Methyl-2-thiophen-2-yl-ethyl)-4-piperidin-4-yl]-N-phenyl-propionamide
Synonyms: N-[1-[1-Methyl-2-(2-thienyl)ethyl]-4-piperidinyl]propionanilide
α-Methylthienylfentanyl

Chemical Structure:

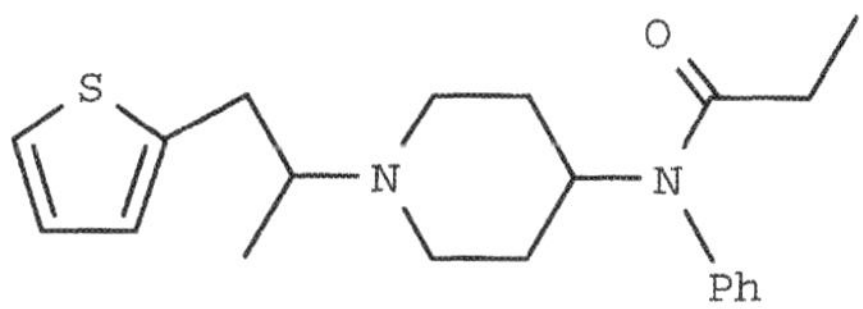

MF: $C_{21}H_{28}N_2OS$
MW: 356.525
CA Registry Number: [103963-66-2]; hydrochloride [117332-94-2]
CA Chemical Name: Propanamide, N-[-1-[1-methyl-2-(2-thienyl)ethyl]-4-piperidinyl]-N-phenyl-
Category: Synthetic opiate
Street Names: China White; Persian White; Mexican Brown; China town; Poison; Tango & Cash; Friend; Goodfellas
Abuse: Limited
Type of action: Euphoriant, analgesic
Human active dose: About 0.1 mg
Duration of action: Not reported
Toxic manifestations: Miosis, nausea, vomiting, orthostatic hypotension, constipation, convulsions, muscular hyperactivity, respiratory depression, pulmonary oedema
Toxicity: Not reported

The drug was detected in California in 1985 [1]. It may be sold under the street names common for all fentanyl analogues. Its identification by C13-NMR [2] has been described. Also, its assay in human urine samples by the RIA based method has been reported [3].

References

[1] Henderson, G.: J. Forensic Sci. **33**, p.569, (1988)
[2] Brine, G.A.; Boldt, K.G.; Huang, P.T.; Sawyer, D.K.; Carroll, F.I.: J. Heterocyclic Chem., **26**, p.677, (1989)
[3] Watts, V.W.; Caplan, Y.H.: J. Anal. Toxicol., **14**, p.266, (1990)

2.9. Metaqualone and its Analogues

Metaqualone (**1**) used to be a very popular non-barbiturate hypnotic but it became a controlled substance because of its strong habit-forming properties [1]. In man, besides its sedative and hypnotic action, it shows anticonvulsant, antitussive and antihistaminic effects. Nausea, vomiting, muscular hyperactivity and respiratory depression are the most frequent signs of a moderate overdose. Clinical management of poisoning due to this type of drug has been described [2].

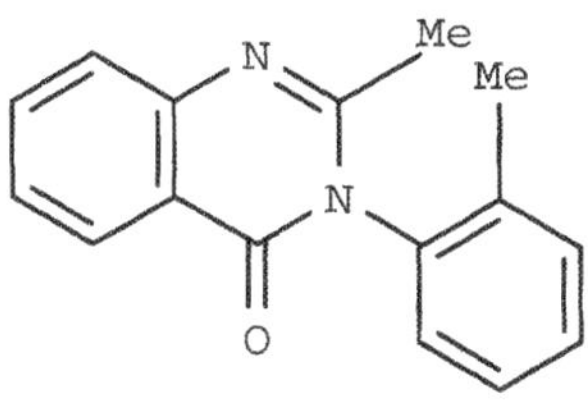

1

100-200 mg

This world-wide abused substance is illegally manufactured in the United States but also in India and Africa [1]. Oral administration of the drug is the most frequent but, being relatively stable, metaqualone and its analogues may also be smoked [1]. Among its abusers these substances are reputed to possess aphrodisiac properties [1].

The key-precursor in its manufacture is N-acetyl anthranilic acid (**2**) which is a chemical widely used in the chemical industry. Attempts to control the commerce of this precursor have brought about limited success [3]. A number of metaqualone analogues may be obtained in a single step operation from the precursor above and a suitable substituted aniline [4]. A seizure of the analogues (**3, 4, 5, 6**) has been reported [1,3,5,6].

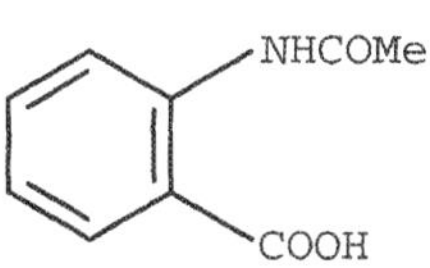

2

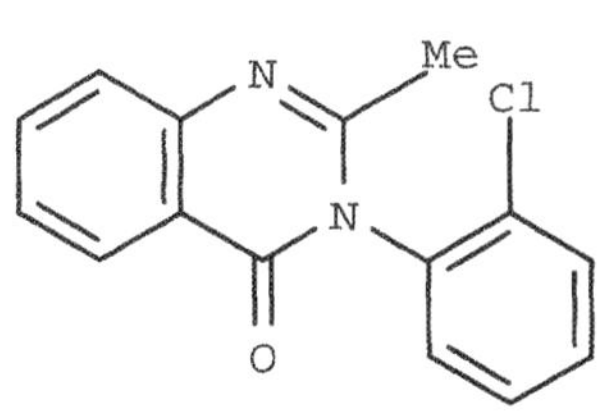

3

100-200 mg

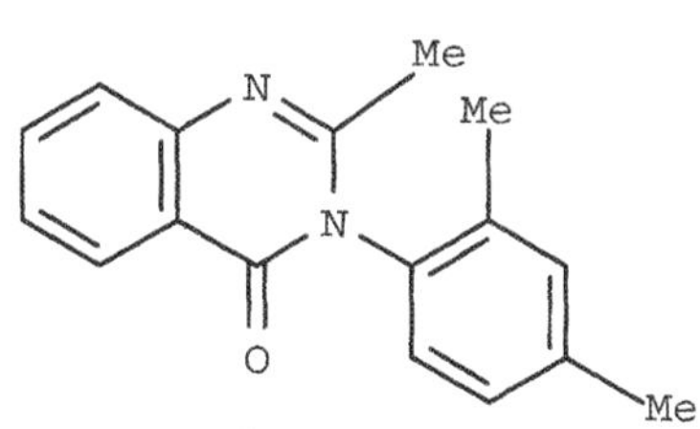

4

100-200 mg

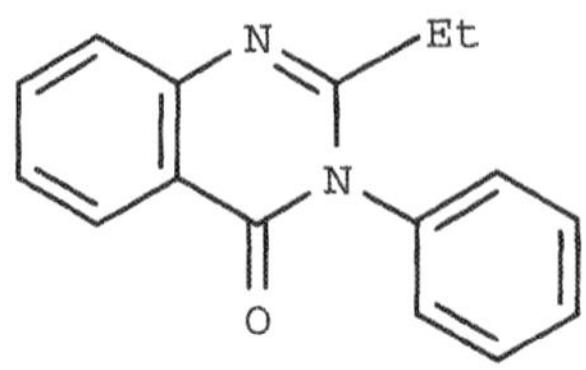

5
100-200 mg

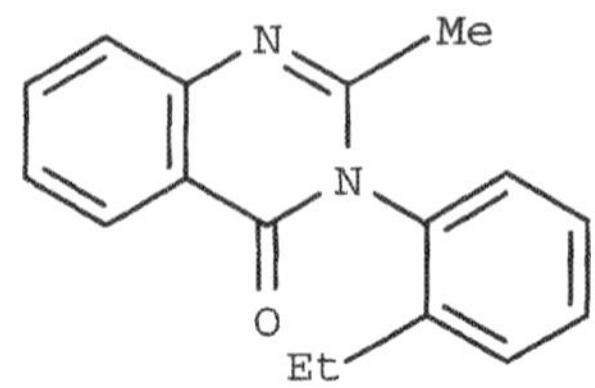

6
100-200 mg

2.9.1 General References

[1] Angelos, S.A.; Meyers, J.A.: J. Forensic Sci., **30**, p.1022, (1985)
[2] Litovitz, T.: Metaqualone. In: Clinical Management of Poisoning and Drug Overdose, Eds. Haddad, L.M.; Winchester, J.F.; W.B. Saunders Company, Eastbourne, UK, p.466, (1983)
[3] Report of International Narcotics Control Board, section « Precursors », (1995)
[4] Grimmel, H.W.; Guenther, A.; Morgan, J.F.: J. Am. Chem. Soc., **68**, p.542, (1946)
[5] Angelos, S.A.: J. Forensic Sci., **38**, p.455, (1993)
[6] Lorimer, P.: Microgram, XVII, p.72, (1984)

2.9.2 Data Sheets: Metaqualone and its Analogues

Substance Number 1 in Chapter 2.9

IUPAC Name: 2-Methyl-3-o-tolyl-3H-quinazolin-4-one
Synonyms: 2-Methyl-3-(2-methylphenyl)-4(3H)-quinazolinone
Metaqualone

Chemical Structure:

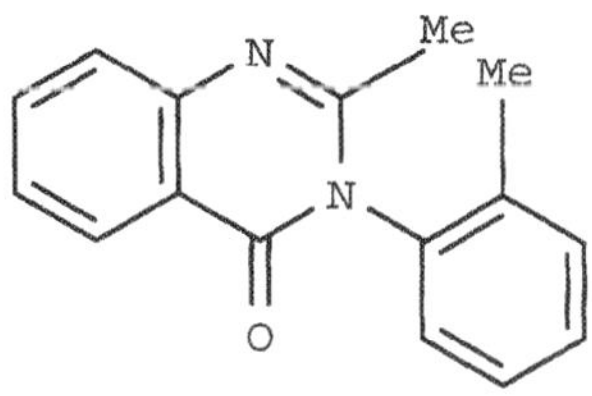

MF: C16H14N2O
MW: 250.299
CA Registry Number: [72-44-6]
CA Chemical Name: 4(3H)-Quinazolinone, 2-methyl-3-(2-methylphenyl)-
Category: Metaqualones
Street Names: Sopors; Sopes; Lude; Heroin for lovers; Love drug
Abuse: Very frequent
Type of action: Hypnotic, euphoriant, aphrodisiac?
Human active dose: 100-200 mg
Duration of action: 5-6 hours
Toxic manifestations: Nausea, vomiting, agranulocytosis, muscular hyperactivity, respiratory depression, coma
Toxicity: LD50 is 255 mg/kg p.o. (rat). In man, the plasmatic concentrations of this substance over 0.1 mg/mL are considered critical.

This drug used to be a widely prescribed hypnotic (Quaalude, Mandrax etc.). After its habit forming properties were discovered, the substance was placed on the list of controlled substances.

At present a number of clandestine laboratories in the USA, India and Africa manufacture large amounts of this drug [1]. The identification of this substance [2,3] and its precursors [2] by GC, HPLC and MS has been described.

References:

[1] Report of International Narcotics Control Board, section « Precursors », (1995)
[2] Angelos, S.A.; Meyers, J.A.: J. Forensic Sci., **30**, p.1022, (1985)
[3] Mills III, T.; Roberson, J.C.; McCurdy, H.H.; Wall, W.H.: Instrumental Data for Drug Analysis, Elsevier Science Publishing Co., Inc., p.1390, (1992)

Substance Number 3 in Chapter 2.9

IUPAC Name: 3-(2-Chloro-phenyl)-2-methyl-3H-quinazolin-4-one
Synonyms: 3-(2-Chlorophenyl)-2-methyl-4(3H)-quinazolinone
Mecloqualone

Chemical Structure:

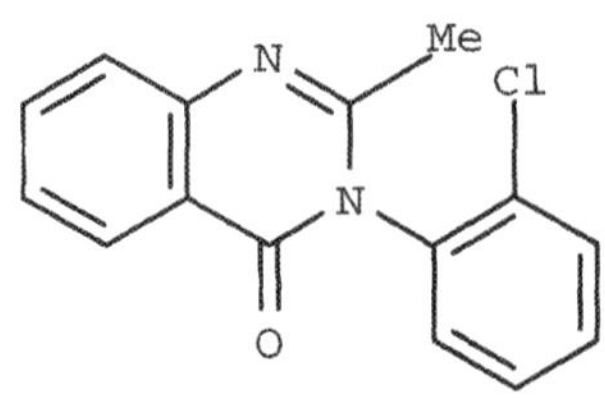

MF: C15H11ClN2O
MW: 270.718
CA Registry Number: [340-57-8]
CA Chemical Name: 4(3H)-Quinazolinone, 3-(2-chlorophenyl)-2-methyl-
Category: Metaqualones
Street Names: Sopors; Sopes; Lude; Heroin for lovers; Love drug
Abuse: Frequent
Type of action: Hypnotic, euphoriant, aphrodisiac?
Human active dose: 100-200 mg
Duration of action: 5-6 hours
Toxic manifestations: Nausea, vomiting, agranulocytosis, muscular hyperactivity, respiratory depression, coma
Toxicity: Not reported

This drug used to be a widely prescribed hypnotic [1] in Europe (« Nubarene »). At present a number of clandestine laboratories manufacture large amounts of this drug from N-acetyl anthranilic acid and o-chloroaniline [1,2]. The identification of this substance [1,2] and its precursors [2] by GC, HPLC and MS has been described.

References:

[1] Angelos, S.A.; Meyers, J.A.: J. Forensic Sci., **30**, p.1022, (1985)
[2] Mills III, T.; Roberson, J.C.; McCurdy, H.H.; Wall, W.H.: Instrumental Data for Drug Analysis, Elsevier Science Publishing Co., Inc., p.1334, (1992)

Substance Number 4 in Chapter 2.9

IUPAC Name: 3-(2,4-Dimethylphenyl)-2-methyl-3H-quinazolin-4-one
Synonyms: 2-Methyl-3-m-xylyl-3H-quinazolin-4-one
3-(2,4-Dimethylphenyl)-2-methyl-4(3H)-quinazolinone

Chemical Structure:

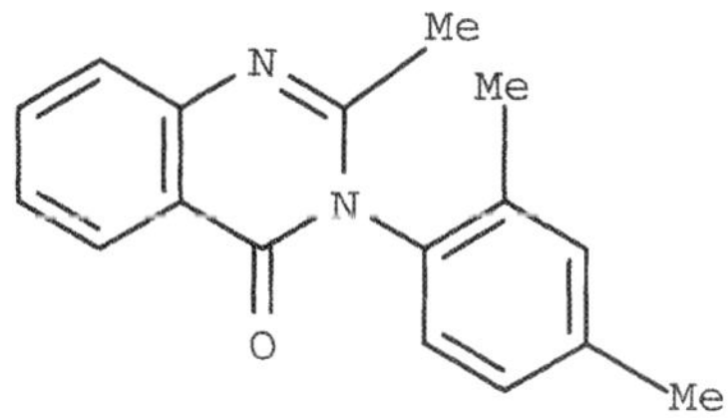

MF: C17H16N2O
MW: 264.326
CA Registry Number: [1915-80-6]
CA Chemical Name: 4(3H)-Quinazolinone, 3-(2,4-dimethylphenyl)-2-methyl-
Category: Metaqualones
Street Names: Sopors; Sopes; Lude; Heroin for lovers; Love drug
Abuse: Rare
Type of action: Hypnotic, euphoriant
Human active dose: 100-200 mg
Duration of action: 5-6 hours
Toxic manifestations: Nausea, vomiting, agranulocytosis, muscular hyperactivity, respiratory depression, coma
Toxicity: Not reported

This designer metaqualone analogue was seized in the USA [1]. The reference also describes the identification of this drug by GC-MS. The substance can be easily prepared from N-acetyl anthranilic acid and 1,3-dimethylaniline.

References:

[1] Angelos, S.A.: J. Forensic Sci., **38**, p.455, (1993)

Substance Number 5 in Chapter 2.9

IUPAC Name: 2-Ethyl-3-phenyl-3H-quinazolin-4-one
Synonyms: 2-Ethyl-3-phenyl-4(3H)-quinazolinone

Chemical Structure:

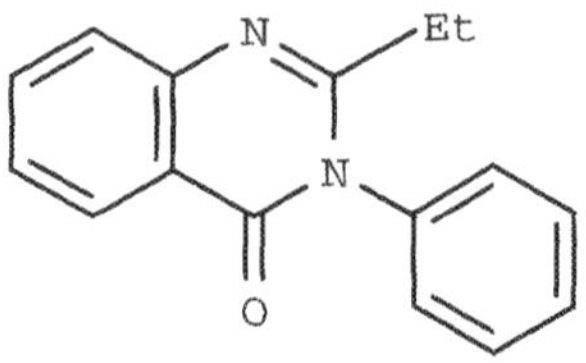

MF: C16H14N2O
MW: 250.299
CA Registry Number: [5260-41-3]
CA Chemical Name: 4(3H)-Quinazolinone, 2-ethyl-3-phenyl-
Category: Metaqualones
Street Names: Sopors; Sopes; Lude; Heroin for lovers; Love drug
Abuse: Rare
Type of action: Hypnotic, euphoriant
Human active dose: 100-200 mg
Duration of action: 5-6 hours
Toxic manifestations: Nausea, vomiting, agranulocytosis, muscular hyperactivity, respiratory depression, coma
Toxicity: Not reported

This designer metaqualone analogue was seized in the USA in 1984 [1,2]. These references also provide MS, IR and NMR spectra of this substance. It can be prepared from N-propionyl anthranilic acid and aniline using the standard metaqualone synthetic procedure. However, propionic anhydride, which is utilised in the synthesis of the first precursor, is a very closely watched chemical, because it is also the necessary precursor for the manufacture of powerful opiates. Hence, the drug is unlikely to reappear in illicit trade today.

References:

[1] Lorimer, P.: Microgram, XVII, p.72, (1984)
[2] Kiser, W.O.: Microgram, XVII, p.75, (1984)

Substance Number 6 in Chapter 2.9

IUPAC Name: 3-(2-Ethyl-phenyl)-2-phenyl-3H-quinazolin-4-one
Synonyms: 3-(2-Ethylphenyl)-2-methyl-4(3H)-quinazolinone
Ethinazone

Chemical Structure:

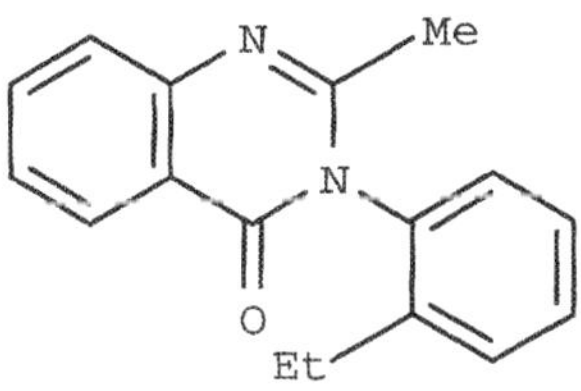

MF: C17H16N2O
MW: 264.326
CA Registry Number: [7432-25-9]
CA Chemical Name: 4(3H)-Quinazolinone, 3-(2-ethylphenyl)-2-methyl-
Category: Metaqualones
Street Names: Sopors; Sopes; Lude; Heroin for lovers; Love drug
Abuse: Rare
Type of action: Hypnotic, euphoriant
Human active dose: 100-200 mg
Duration of action: 5-6 hours
Toxic manifestations: Nausca, vomiting, agranulocytosis, muscular hyperactivity, respiratory depression, coma
Toxicity: Not reported

This drug used to be a widely prescribed hypnotic and sedative in Europe (« Aolan »). At present, it is occasionally manufactured in clandestine laboratories [1]. The necessary precursor, o-ethylaniline, is a readily available chemical. The identification of this drug by GC-MS and HPLC has been described [2].

References:

[1] Report of International Narcotics Control Board, section « Precursors », (1995)
[2] Mills III, T.; Roberson, J.C.; McCurdy, H.H.; Wall, W.H.: Instrumental Data for Drug Analysis, Elsevier Science Publishing Co., Inc., p.860, (1992)

2.10 GHB

This drug is simple sodium (or more rarely potassium) salt of γ-hydroxybutyric acid (**1**)[1,2].

It can be very easily prepared from γ-butyrolactone (**2**) by alkaline hydrolysis. The reaction mixture, after eventual pH correction, is then directly sold to consumers.

1

1-3 g

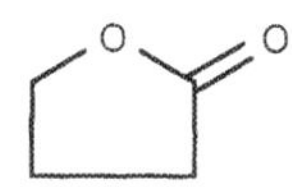

2

This drug is usually sold under street names such as « Liquid E », « Liquid X » and « Liquid Ecstasy » at « Rave » parties [2].

The drug shows an inhibition removing activity in humans which, in some respects, seems to resemble that of MDMA, but in contrast to this amphetamine, the substance produces hypotension and ataxia [3].

Several lethal cases of intoxication involving GHB with other substances of abuse have been reported [4].

2.10.1 General References

[1] Robson, P.: Forbidden Drugs, Oxford University Press, p.84, (1994)
[2] http://ceida.net.au/latest/fantasy.html
[3] Stephens, B.G. et al.: J. Anal. Toxicol., **18**, p.357, (1994)
[4] Ferrara, S.F. et al.: J. Forensic Sci., **40**, p.501, (1995)

2.10.2 Data Sheets: GHB

Substance Number 1 in Chapter 2.10

IUPAC Name: Sodium (potassium) 4-hydroxybutyrate
Synonyms: Sodium (potassium) γ-hydroxybutyrate
Sodium Oxybate

Chemical Structure:

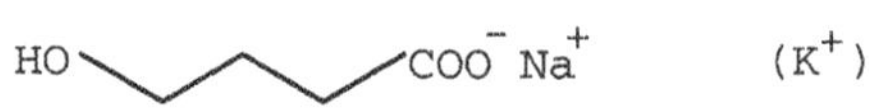

MF: C4H7O2 Na; C4H7O2 K
MW: Na salt: 126.087; K salt: 142.199
CA Registry Number: Acid [591-81-1]; sodium salt [502-85-2]; potassium salt [57769-01-4]
CA Chemical Name: Butanoic acid, 4-hydroxy- sodium or potassium salt
Category: GHB
Street Names: GHB; Liquid E; Liquid X; Liquid Ecstasy; Fantasy; GBH
Abuse: Frequent
Type of action: Tranquiliser, euphoriant, entactogen
Human active dose: 1-3 g
Duration of action: 2-4 hours
Toxic manifestations: Hypotension, bradycardia, ataxia, coma
Toxicity: In man, administration of this drug at doses higher than 50 mg/kg may rapidly lead to loss of consciousness.

This drug is sold under the street names indicated above at « Rave » parties [1]. Among its abusers the substance is believed to possess an aphrodisiac effect. Being reputed to induce a marked increase in muscular mass, this substance is also frequently abused by body-builders [2,3]. The last reference also describes the identification of this drug by IR and GC-MS. In some countries (e.g. Italy) the drug has been accepted in treatment of alcohol withdrawal syndrome [4].

References:

[1] http://www.erowid.com/entheogens/ghb/ghb_media2.shtml
[2] http://www.altculture.com/site/entries/ghb.html
[3] Blackledge, R.D.; Miller, M.D.: Microgram, XXIV, p.172, (1991)
[4] Gallimberti, L. et al.: The Lancet, Vol. **2**, No.8666, p.787, (1989)

Subject Index

A

B

E

F

G

H

I

J

K

L

M

N

O

P

Q

R

S

T

Street Names Index

Abbreviations

AL	4-Allyloxy-3,5-dimethoxyphenethylamine
ALPHA	1-(3,4-Methylenedioxyphenyl)-1-propanamine
BOH	b-Methoxy-homopiperonylamine
BZ	3-Quinuclidinyl benzilate
CAT	Norephedrone
2CB	4-Bromo-2,5-dimethoxyphenethylamine
2CC	4-Chloro-2,5-dimethoxyphenethylamine
2CI	4-Iodo-2,5-dimethoxyphenethylamine
2CN	2,5-Dimethoxy-4-nitrophenethylamine
CNS	Central Nervous System
2CT	2,5-Dimethoxy-4-methylthiophenethylamine
DET	N,N-Diethyltryptamine
DIPT	N,N-Diisopropyltryptamine
DMA	2,5-Dimethoxyamphetamine
DMT	N,N-Dimethyltryptamine
DOB	4-Bromo-2,5-dimethoxyamphetamine
DOC	4-Chloro-2,5-dimethoxyamphetamine
DOET	4-Ethyl-2,5-dimethoxyamphetamine
DOI	4-Iodo-2,5-dimethoxyamphetamine
DOM	2,5-Dimethoxy-4-methylampahetamine
DOPR	2,5-Dimethoxy-4-propylamphetamine
DOT	2,5-Dimethoxy-4-methylthioamphetamine
DPT	N,N-Dipropyltryptamine
EA	4-Ethoxyamphetamine
ELISA	Enzyme Linked Immunosorbent Assay
EMA	4-Ethoxy-3-methoxyamphetamine
ET	Etryptamine
GAMMA	N-Methyl-1-(3,4-methylenedioxyphenyl)-3-propanamine
GHB	g-hydroxybutyric acid
HMDA	1-(3,4-Methylenedioxyphenyl)-3-butanamine
HMDMA	N-Methyl-1-(3,4-methylenedioxyphenyl)-2-butanamine
IMAO	Inhibitors of Monoaminooxidase
JB-336	1-Methyl-3-piperidyl benzilate
JB-840	1-Methyl-3-piperidyl cyclohexyl phenyl glycolate
LAMPA	D-Lysergic acid methylpropylamide
LSM	D-Lysergic acid morpholide
MALPHA	N-Methyl-1-(3,4-methylenedioxyphenyl)-1-propanamine
MBDB	N-Methyl-1-(3,4-methylenedioxyphenyl)-2-butanamine
MDA	3,4-Methylenedioxyamphetamine
MDAOH	N-Hydroxy-3,4-methylenedioxyamphetamine
MDBA	1-(3,4-Methylenedioxyphenyl)-2-butanamine
MDDM	N,N-Dimethyl-3,4-methylenedioxyamphetamine
MDEA	N-Ethyl-3,4-methylenedioxyamphetamine
MDMA	3,4-Methylenedioxymethamphetamine
MDMAOH	N-Hydroxy-3,4-methylenedioxymethamphetamine
MDPR	3,4-Methylenedioxy-N-propylamphetamine

MEM	4-Ethoxy-2,5-dimethoxyamphetamine
5MEODMT	5-Methoxy-N,N-dimethyltryptamine
MEPEA	4-Ethoxy-3-methoxyphenethylamine
MMDA	3-Methoxy-4,5-methylenedioxyamphetamine
2MMDA	2-Methoxy-3,4-methylenedioxyamphetamine
MPTP	1,2,5,6,-Tetrahydro-1-methyl-4-phenylpyridine
OPPPP	1-(3-Oxo-3-phenylpropyl)-4-phenyl-4-piperidinol propionate
PEPAP	1-Methyl-4-phenyl-4-piperidyl propionate
PCE	N-Ethyl-1-phenylcyclohexylamine
PCP	Phencyclidine
PHP	1-(1-Phenylcyclohexyl)pyrrolidine
4MA	4-Methoxyamphetamine
RIA	Radioimmunnoassay
4MMA	4-Methoxymethamphetamine
T2	4-Ethylthio-2,5-dimethoxyphenethylamine
TCP	1-[1-(2-Thienyl)cyclohexyl]piperidine
TCPy	1-[1-(2-Thienyl)cyclohexyl]pyrrolididine
D3-THC	Tetrahydro-6,6,9-trimethyl-3-pentyl-6H-dibenzo[b,d]pyran-1-ol
TMA	3,4,5-Trimethoxyamphetamine
TMA2	2,4,5-Trimethoxyamphetamine
5TOM	2-Methoxy-4-methyl-5-thiomethylamphetamine

Glossary

Aesthetic enhancer:	Usually a low dose of a psychedelic that enhances enjoyment of various artistic events
Agranulocytosis:	A serious illness due to an absence of granulocytes in the blood
Anaesthesia:	Partial or total inability to feel all forms of sensations (pain, heat, cold etc.)
Analogue:	A chemical compound that is more or less structurally modified and related to a controlled substance
Anorexia:	Loss of appetite
Anticholinergic:	Agent, capable of counteracting the action of acetylcholine
Antidepressant:	A drug used to treat mental depression
Ataxia:	Loss of motor co-ordination
Booster:	A small dose of a drug taken to prolong the action of the same drug
Bradycardia:	An abnormally slow heart rate
Catalepsy:	Prolonged pathological maintenance of physical attitudes, often characterised by a particular rigidity of the body (« frozen attitude »)
Catatonia:	A condition of marked alteration in muscle tone and/or motor reactivity (posturing, rigidity) frequently seen in schizophrenia
Coma:	Unconscious state due to illness or intoxication
Confusion:	A state of disoriented, clouded mind
Convulsions:	Violent involuntary movement of the body caused by chaotic muscular contractions
Deliriant:	A drug capable of inducing a state of delirium
Delirium:	Usually wild excitement, accompanied with clouded consciousness, impaired orientation, fragmented thought and speech and hallucinations
Delusions:	A false belief or opinion, contrary to reality
Depersonalisation:	Feeling of loss of personal identity
Derealisation:	Feeling that external reality is suddenly unfamiliar or unreal
Eidetic:	Expression approximately synonymous to the term hallucinogen
Empathogen:	An empathy inducing drug
Entactogen:	A drug capable of improving communicative contact among participants
Entheogen:	A recently proposed term to replace the term « psychedelic » for ritual purposes
Euphoriant:	A mood enhancing drug (« mood enhancer »)
Hallucinations:	Seeming to see, hear, smell, taste or touch something (or someone) which is not really present
Hallucinogen:	A drug capable inducing hallucinations
Hepatotoxic:	Toxic to the liver
Hyperreflexia:	Abnormal increase of muscular reflexes
Hypertension:	High blood pressure

Hypotension:	Low blood pressure
Illusions:	A false or misinterpreted perception of reality
Incapacitating agent:	A chemical warfare agent capable of totally disrupting one's normal behaviour
Miosis:	Marked constriction of the eye pupil
Mood enhancer:	See Euphoriant
Mydriasis:	Marked dilatation of the eye pupil
Precursors:	Key chemicals, necessary to manufacture a given designer drug
Primer, priming:	A primer is a substance, showing practically no psychotomimetic properties in man, which is taken before an active psychotomimetic ("primed drug") to enhance and/or modify its action. Some Internet sources also utilise this term for similar combinations of two active psychotomimetics.
Psychedelic:	Meaning « mind manifestation » inducing agent
Psychodysleptic:	Meaning « mind disrupting » agent
Psychotic decompensation:	Reappearance and worsening of previously stabilised mental disease (psychosis)
Psychotogen:	Meaning « psychosis inducing » agent
Psychotomimetic:	Meaning « psychosis mimicking » agent
Synesthesias:	Perceptional disturbance when one stimulus induces a secondary sensation involving one of the other senses. Thus, sounds may have colours etc.
Tachycardia:	Accelerated heart rate
Tremor:	A rhythmic involuntary movement of the body

www.ingramcontent.com/pod-product-compliance
Ingram Content Group UK Ltd.
Pitfield, Milton Keynes, MK11 3LW, UK
UKHW020814290926
15470UKWH00003B/1146